GEOGRAPHIC INFORMATION SYSTEMS
An Introduction

JEFFREY STAR
JOHN ESTES

University of California, Santa Barbara

PRENTICE HALL, Englewood Cliffs, New Jersey 07632

Library of Congress Cataloging-in-Publication Data

Star, Jeffrey.
Geographic information systems : an introduction / Jeffrey Star,
John Estes.
p. cm.
Bibliography: p.
Includes index.
ISBN 0-13-351123-5
1. Geography—Data processing. I. Estes, J. E. II. Title.
G70.2.S73 1990 89-33220
910′ .285—dc20 CIP

Editorial/production supervision: Christina Ferrari
Cover design: Photo Plus Art
Manufacturing buyer: Paula Massenaro
Cover photograph: California Condor database. Photo by Jeffrey Star.
Color insert layout: Thomas Nery

A Division of Simon & Schuster
Englewood Cliffs, New Jersey 07632

Printed in the United States of America

10 9 8 7 6 5 4 3

ISBN 0-13-351123-5

Prentice-Hall International (UK) Limited, *London*
Prentice-Hall of Australia Pty. Limited, *Sydney*
Prentice-Hall Canada Inc., *Toronto*
Prentice-Hall Hispanoamericana, S.A., *Mexico*
Prentice-Hall of India Private Limited, *New Delhi*
Prentice-Hall of Japan, Inc., *Tokyo*
Simon & Schuster Asia Pte. Ltd., *Singapore*
Editora Prentice-Hall do Brasil, Ltda., *Rio de Janeiro*

To Toni and Claire

Brief Contents

1 Introduction 1

2 Background and History 14

3 The Essential Elements of a GIS: An Overview 24

4 Data Structures 32

5 Data Acquisition 61

6 Preprocessing 76

7 Data Management 126

8 Manipulation and Analysis 143

9 Product Generation 174

10 Remote Sensing and GIS 191

11 Practical Matters 220

12 Applications 230

13 Looking Toward the Future 245

Suggestions for Exercises 252
Glossary and Abbreviations 261
Bibliography 283
Index 297

Contents

Preface xi

1 Introduction 1

1.1 Geography 1
1.2 Information Systems 2
1.3 A Manual Geographic Information System 4
1.4 Applications 8
1.5 Geographical Concepts 9
1.6 The Four Ms 12

2 Background and History 14

2.1 The Cartographic Process 15
2.2 Early History 17
2.3 The Modern Era 19

3 The Essential Elements of a GIS: An Overview 24

3.1 Functional Elements 24
3.2 Data in a GIS 28

4 Data Structures 32

4.1 Raster Data Structures 33
4.2 Vector Data Structures 48
4.3 Comparisons between Data Structures 57

5 Data Acquisition 61

5.1 Introduction 61
5.2 Existing Datasets 64
5.3 Developing your own Data 70

6 Preprocessing 76

6.1 Format Conversion 77
6.2 Data Reduction and Generalization 91
6.3 Error Detection and Editing 93
6.4 Merging 95
6.5 Edge Matching 96
6.6 Rectification and Registration 97
6.7 Interpolation 114
6.8 Photointerpretation 118

7 Data Management 126

7.1 Basic Principles of Data Management 127
7.2 Efficiency 130
7.3 Conventional Database Management Systems 134
7.4 Spatial Database Management 139

8 Manipulation and Analysis 143

8.1 Reclassification and Aggregation 144
8.2 Geometric Operations on Spatial Data 154
8.3 Centroid Determination 156
8.4 Data Structure Conversion 157
8.5 Spatial Operations 157
8.6 Measurement 164
8.7 Statistical Analysis 167
8.8 Modeling 169

9 **Product Generation** **174**

9.1 Types of Output Products 174
9.2 Hardware Components 183

10 **Remote Sensing and GIS** **191**

10.1 A Brief History 193
10.2 Remote Sensing Technology 195
10.3 Digital Processing of Remotely Sensed Data 209
10.4 Interfacing Remote Sensing and GIS's 213
10.5 Synergism 215

11 **Practical Matters** **220**

11.1 A Case Study 220
11.2 Database Volume 223
11.3 Specifications 225
11.4 Project Management 228

12 **Applications** **230**

12.1 Master Planning 230
12.2 Proposed Dam Site 232
12.3 Waste Site Selection 233
12.4 Irrigation and Water Resource Potential 234
12.5 Merging Raster and Vector Data for Map Update 235
12.6 Species Habitat Analysis 236
12.7 Municipal GIS 238
12.8 Agricultural Production Modeling 240
12.9 KBGIS: Artificial Intelligence and GIS 242

13 **Looking Toward the Future** **245**

Suggestions for Exercises 252
Glossary and Abbreviations 261
Bibliography 283
Index 297

Preface

This book was written for several reasons. First, we hope it helps to fill a gap by providing an entry into the literature of geographic information systems. Up until very recently, most of the information about geographic information systems (GISs) has been in the fugitive, or *gray*, literature; the latter includes in-house technical reports, conference proceedings, non-refereed publications, and poorly distributed manuscripts describing efforts funded by federal and state agencies. Publications from professional conferences, such as Purdue University's "Machine Processing of Remotely Sensed Data" Symposia, the Harvard Computer Graphics Symposia, and the Urban and Regional Information Science Association meetings have been some of the best places to find current information about GISs, in terms of both the theoretical and the practical aspects of the field.

Special-purpose documents have also been useful sources, for those of us who know where to look. Briefing documents prepared by such agencies as the U.S. Forest Service and companies such as the Environmental Systems Research Institute, and surveys of the field by the American Farmland Trust, have been valuable resources in their respective narrow domains. Publications such as the American Society of Photogrammetry and Remote Sensing's *Photogrammetric Engineering and Remote Sensing* also have articles about GISs on a regular basis. Recently, journals have been started that emphasize spatial data processing. Two of these new journals are the *International Journal of Geographical Information Systems*, published quarterly by Taylor and Francis in the U.K., and *Aerial and Space Imaging, Remote Sensing and Integrated Geographical Systems*, published quarterly by Imaging Systems Publications in Sweden. Recent conferences have focused specifically on geographic information systems. In particular, the proceedings of GIS '87 (October, 1987; San Francisco) and IGIS '87 (November, 1987; Crystal City, Virginia) provide a summary of the state of the field, although primarily from an academician's point of view.

Information about geographic information systems may sometimes be found in one chapter of a book - see, for example, Short's *Landsat Tutorial Workbook* (1982), a classic volume which is unfortunately out of print, or the chapter edited by Marble in the second edition of the *Manual of Remote Sensing* (American Society of Photogrammetry and Remote Sensing, 1983). Recently, a monograph has been published that focuses completely on GIS's. Burrough's *Principles of Geographical Information Systems for Earth Resources Assessment* (1986) is an excellent resource in this rapidly emerging field. For information on a variety of aspects of GIS, see Marble, Calkins, and Peuquet (1984) - a self-published three ring binder of readings on GIS that has also become a well-known standard reference.

A second important function for this text is to serve as an introduction to the field, but at the upper-division undergraduate or beginning graduate level. Much of the material that follows has been used in short courses we have presented under the auspices of George Washington University and elsewhere. We hope it serves as a text for a one-semester course focusing on geographic information systems and their applications, as well as a resource for individual study. A student with some background in computer graphics, cartography, and geographic theory should find this text accessible.

A final reason for this book's existence is to raise the banner of *data integration* as a philosophical basis for modern GIS's. From one perspective, the essence of a geographic information system is its integration of the many different kinds of information we may be able to obtain about the spatial objects in an area. In other words, a user of a GIS should be able to work with many different types of data, bringing *all* the information relevant to a problem together in a consistent form, and should be able to bring to bear on the problem all the power of all the sophisticated analysis tools available. By the same token, we believe a geographic information system is the correct tool for integrating the different technologies that are used in gathering, analyzing, and assessing spatial data.

We begin with a chapter that introduces some of the jargon of geographic information systems. It does so by presenting a completely non-digital land-use planning problem: the design of a golf course. This is followed by a chapter on the history of geographic information systems, and by another that introduces the essential components of a GIS. Chapter 4 discusses data structures: how a GIS actually stores the kinds of data and information that are commonly used in systems.

In the next five chapters we explore the five essential parts of a GIS: data acquisition, preprocessing, database management, analysis and manipulation, and producing the final output. In these chapters we attempt to show the practical use of the different tools and techniques.

Chapter 10 examines the relationship between the technology of remote sensing and the philosophy of geographic information systems. From a theoretical point of view, the image processing systems used when working with remotely sensed data are geographic information systems, although with strengths in certain kinds of numerical operations and presentation functions. Based on our experience, we believe that remotely sensed data and GIS's are natural allies. In this chapter we have tried to highlight the interdependencies of these two areas, and indicate the potential each has for the other.

Chapter 11 briefly discusses some practical matters. Bringing a digital geographic information system into an organization requires changing the way the organization conducts its day-to-day operations. This affects an organization's management in several ways, including short-term disruptions and costs during the transition to a digital system, as well as changes to the long-term planning process. A GIS makes new demands on staff and staff development, and affects funding requirements in both the short and long term. Some of these are the same as the impacts of any other digital information system, and thus, there are some relevant guidelines and case studies. Problems specific to a GIS, in terms of estimating database volume, developing specifications, and managing the overall project, are discussed briefly.

In Chapter 12 we present a potpourri of applications, drawn from friends and colleagues who were able to contribute material to us. These applications cover a wide range, from the design and management of large facilities to long-range research on the global carbon cycle. The applications also touch on the wide range of hardware and capabilities now used to solve spatial data processing problems.

Finally, in the last chapter we gather our thoughts about the future. Shakespeare's Macbeth had perfect counsel about future events from three witches. We claim no such accuracy, and hope our prophecies are less disastrous. However, there are visible trends in the GIS industry, as well as interesting new developments coming out of the research laboratories. New data acquisition technologies are being developed and data continues to be less expensive to process and store, while our appetite for processing continues to grow. We report on some of these developments, both to describe the current state of the art and to show some of the directions we believe the state of the practice will take.

Several color plates are provided to help illustrate specific topics presented in the text. They are:

1. Aerial photograph: color infrared
2. Master planning
3. Proposed dam site placement
4. Waste site selection
5. African water resources
6. Raster/vector data for map update
7. California Condor database
8. Agricultural production modeling

For contributions of applications and figures, we thank John Jensen (University of South Carolina), Bruce Rado and Lawrie Jordan (ERDAS Inc., Atlanta, Georgia), S.J. Camarata (Environmental Systems Research Institute Inc., Redlands, California), and Heidi Schweikart (Intergraph Corp., Huntsville, Alabama).

We gratefully acknowledge the help and encouragement of many of our colleagues - in particular, Larry Carver, Frank Davis, Ellen Knapp, Terry Smith, Larry Tinney, and Waldo Tobler. Several of our students and staff - particularly Key Ho Bahk, Ken McGwire, Terri Morris, Jeff Sandberg, and Joe Scepan - have made valuable contributions to this effort. We particularly thank Hugh Calkins, Dave Cowen, and John Jensen for their careful reading of portions of the manuscript. All remaining errors, of course, are ours alone.

About the Authors

Jeffrey L. Star is a researcher and science manager in the Geography Department at the University of California, Santa Barbara. His Ph.D. is from the Scripps Institution of Oceanography of the University of California. As a lecturer he has taught many students at both undergraduate and graduate levels. In addition he is widely known for intensive workshops in GIS for agencies such as the Agency for International Development, the Defense Mapping Agency Aerospace Center, George Washington University, and the Stennis Space Center. He has served as a consultant and advisor to numerous organizations both public and private, including the United Nations, TRW, NASA, and the National Bureau of Standards.

John E. Estes is professor of Geography and Director of the Remote Sensing Research Unit, University of California, Santa Barbara. He received his Ph.D. from the University of California, Los Angeles. Included in a varied career are duties as Co-Director of the National Center for Geographic Information and Analysis, Visiting Senior Scientist at NASA Headquarters, and numerous roles as consultant to industry. He is a member of the National Academy of Science's Mapping Sciences Committee. His research has been funded by NASA, the U.S. Geological Survey, and the U.S. Department of Agriculture, among others.

Chapter 1

Introduction

As in any other technical area, there can be a fair amount of technical vocabulary to learn before a student can be comfortable with the subject. Since this text is introductory by design, we will try to be consistent in our use of language. Much of the vocabulary of geographic information systems overlaps that of computer science and mathematics in general, and computer graphics applications in particular. We provide a glossary of technical terms at the end of this text, as a reference for the student.

1.1 Geography

Geography has been facetiously defined as that discipline which, when some use is found for it, is called something else. Slightly more serious scholars have defined geography as "what geographers do". The German philosopher Immanual Kant set geography in the context of the sciences by stating that knowledge could be subdivided into three general areas:

1. those disciplines that study particular objects or sets of objects and phenomena (such as biology, botany, forestry, and geology);

2. those disciplines that look at things through time (in particular, history); and

3. those disciplines that look at features within their spatial context (specifically, *geographic disciplines*).

In a more classical sense, the word **geography** may be defined in terms of its constituent parts: *geo* and *graphy*. *Geo* refers to the Earth, and *graphy* indicates a process of writing; thus *geography* (in this literal interpretation) means writing about the Earth.

Another definition of geography focuses on man's relationship with the land. In their writings, geographers deal with spatial relationships. A key tool in studying these spatial relationships is the map. Maps present a graphic portrait of spatial relationships and phenomena over the Earth, whether a small segment of it or the entire globe.

It is interesting that in a survey conducted to determine what factors influenced people to adopt the profession of geography, an early interest in maps rated at the top of the list. There are many skills that people possess to a greater or lesser degree. If a person speaks well, he or she possesses *fluency*. If a person understands writing well, he or she possesses *literacy*. If a person understands numbers and quantitative concepts well, he or she possesses (at least in Great Britain) *numeracy*. Similarly, there is a special skill in the analysis of spatial patterns in two and three dimensions. This skill can be referred to as *graphacy*. Although many individuals take this skill for granted, we all know those who have difficulty reading maps or interpreting aerial photographs. What these two activities have in common is the use of an essentially two dimensional view of geographic space, a view that helps the adept map-reader or photointerpreter to understand spatial relationships.

1.2 Information Systems

The function of an information system is to improve one's ability to make decisions. An **information system** is that chain of operations that takes us from planning the observation and collection of data, to storage and analysis of the data, to the use of the derived information in some decision-making process (Calkins and Tomlinson, 1977). This brings us to an important concept: a map is a kind of information system. A map is a collection of stored, analyzed data, and information derived from this collection is used in making decisions. To be useful, a map must be able to convey information in a clear, unambiguous fashion, to its intended users.

A **geographic information system** (GIS) is an information system that is designed to work with data referenced by spatial or geographic coordinates. In other words, a GIS is both a database system with specific capabilities for

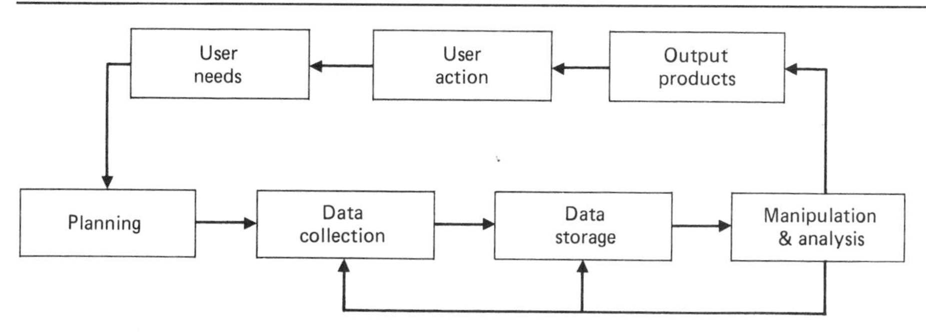

Figure 1.1 Simplified information system overview.

spatially-referenced data, as well a set of operations for working with the data (see Figure 1.1). In a sense, a GIS may be thought of as a higher-order map.

As we shall see later, a modern GIS also stores and manipulates non-spatial data. Just as we have maps designed for specific tasks and users (road maps, weather maps, vegetation maps, and so forth), we can have GISs designed for specific users. The better we are able to understand the range of needs of a user, the better we will be able to provide the correct data and tools to that user.

A geographic information system can, of course, be either **manual** (sometimes called analog) or **automated** (that is, based on a digital computer). Manual geographic information systems usually comprise several data elements including maps, sheets of transparent materials used as overlays, aerial and ground photographs, statistical reports and field survey reports. These sets of data are compiled and analyzed with such instruments as stereo viewers, transfer scopes of various kinds, and mechanical and electronic planimeters. Calkins and Tomlinson (1977) point out that manual techniques could provide the same information as computer-aided techniques, and that the same processing sequences may occur. While this may no longer be entirely true, manual GISs have played an extremely important role in resource management and planning activities. Furthermore, there are still applications where a manual GIS approach is entirely appropriate. Although this text focuses on the technology, instrumentation, and utilization of geographic information systems that are automated, it is still helpful to examine a manual GIS first.

1.3 A Manual Geographic Information System

To introduce some of the language of geographic information systems with a simple first example, let's examine an application of a simple manual GIS. This GIS arises during the early steps in developing a site for a golf course. We assume for this discussion that a specific site is already under consideration. A planner has sought out and gathered together a group of existing datasets for the site. This group might include a topographic map, a blue-line map of parcel boundaries from the local municipal planning agency, and an aerial photograph of the site (Figure 1.2). We refer to these three datasets - 2 maps and a photograph - as **data layers** or **data planes.**

The topographic map depicts several kinds of information. Elevation on the site is portrayed as a series of contour lines. These contour lines provide us with a limited amount of information about the shape of the terrain. Certain kinds of land cover are indicated by colors (often blue for water, green for vegetation) and textures or patterns (such as repeated patterns denoting wetlands). A number of kinds of man-made features are indicated, including structures and roadways, typically by lines and shapes printed in black. In many cases, the information on this map is five to fifteen years out of date, a common situation resulting from the rate of change of land cover in the area and the cycle of map updates. Each of these different kinds of information, which we may decide to store in various ways, is called a **theme**.

The map from the local planning agency provides us with additional and different kinds of information about the area. This map focuses principally on the infrastructure: legal descriptions of the proposed golf course property boundaries, existing and planned roadways, easements of different kinds, and the locations of existing and planned utilities such as potable water, electric and gas supplies, and the sanitary sewer system. The planning map is probably not at the same scale as the topographic map; the former is probably drawn at a larger scale than the latter. Furthermore, the two aren't necessarily based on the same map projection (see section 6.6.1). For a small area like our golf course, the approximate scale of the data is probably more important than the details of the map projection.

The aerial photograph is a rich source of data, particularly for an analyst with some background in image interpretation. A skilled interpreter may be able to detect patterns in soils, vegetation, topography, and drainage, based on the content of the photograph. Unfortunately, this photograph is

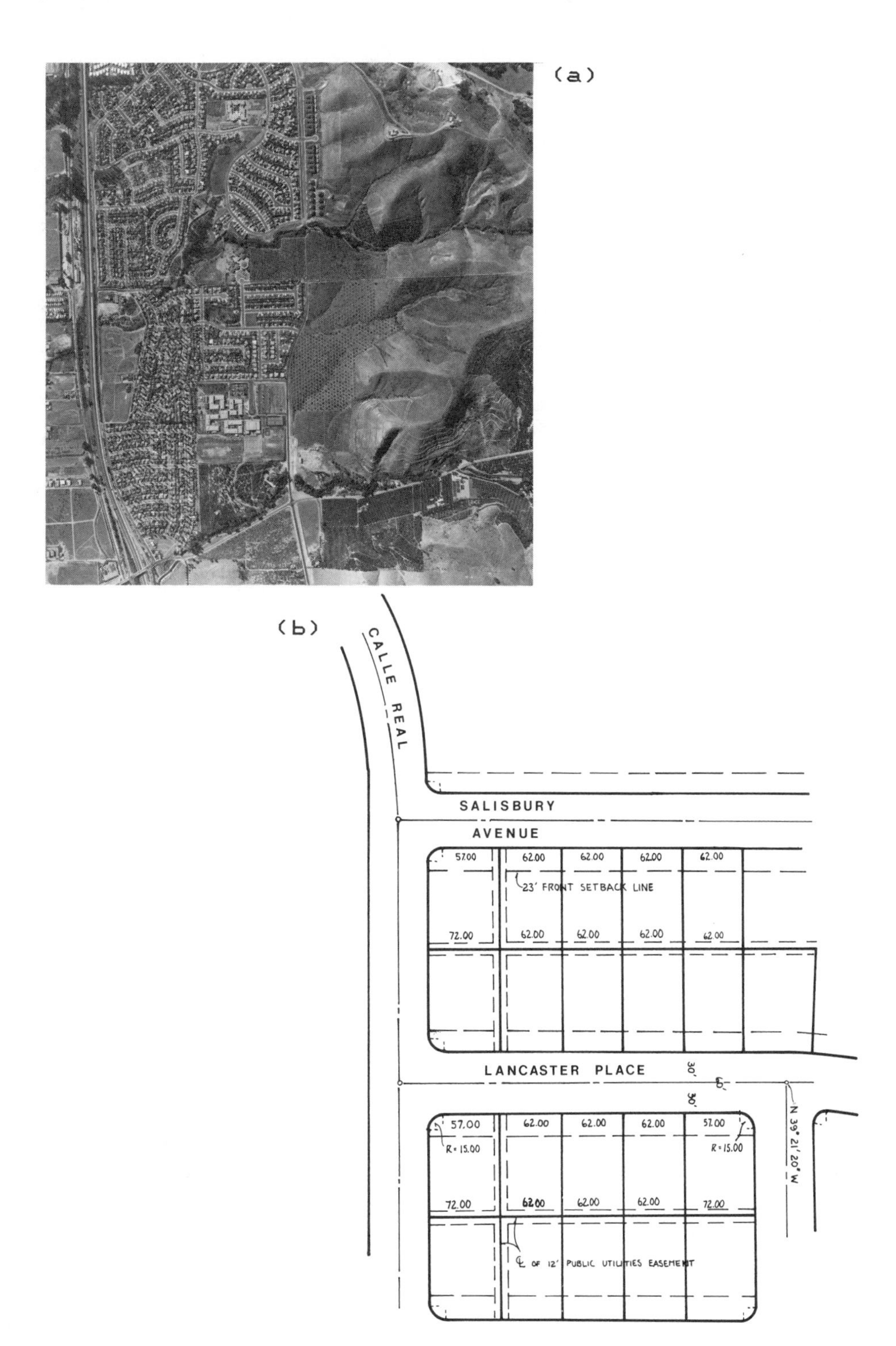

Figure 1.2 A manual GIS. Three different data layers are used to develop a golf course.

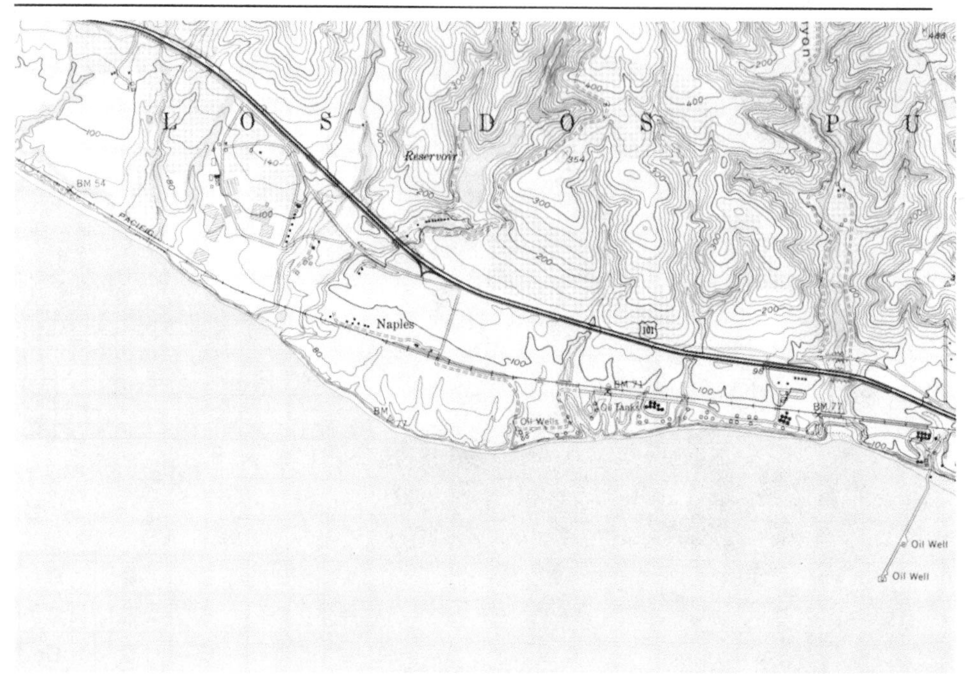

(c)

Figure 1.2 (continued)

probably of different scale than either of the two maps, and may have significant geometric distortions. The two maps attempt to be **planimetric**, that is, the horizontal spatial relationships between objects on the ground are correctly represented on the maps. The photograph, on the other hand, probably suffers from both the perspective distortion inherent in all photographs and from a non-vertical point of view.

A second step in developing plans for this site is to manipulate the three datasets so they can be used simultaneously. A cartographer or draftsperson is given the task of redrawing the municipal planning map and the topographic map onto plastic film, in such a way that the features on the new film-maps overlay their counterparts on the aerial photograph. This process, called **registration**, in effect causes the objects (buildings, roadways, and so forth) to move form their original locations in the planning map, so that they fall at the positions they are found in the photograph. Alternatively, the photograph could be manipulated in such a way that the visible features overlay the corresponding elements on the planning map, and then a transparency could be made. In any case, this spatially registered set of data planes is now a useful geographic **database**. Since the three sets of information have now been converted to overlay each other, further manipulations are much easier. Note that if the original aerial photography were chosen as the base data layer, the resulting database may have no simple relationship to a well-known geodetic coordinate system, such as latitude and longitude. However, for applications that cover a small area, this may not be a serious problem.

Once the individual data layers have been adjusted to a common view of the Earth's surface, there are a number of analytic operations we might make with this manual geographic information system. The analyst begins by drawing some new features on another sheet of plastic that overlays the other data layers. For example, we might generate 25-meter-wide corridors at the edges of the property, and 10-meter-wide corridors around the existing roads and proposed utility locations. These newly derived regions might suggest some places that are unsuitable for development of large new facilities, and others that are particularly desirable due to proximity to needed utilities. As such, we might now be in a position to make preliminary decisions on the location of the club house, storage yards, access roads, and parking facilities.

Next, the planner lays a coarse grid over the database, and start marking the conjunction of topographic and hydrologic features and vegetation that are most suitable for fairways and tee-off areas. Existing waterways could be used as boundaries between areas on the golf course, while the locations of wooded areas could be considered as part of the course plan. Based on these preliminary decisions, we prepare a new data layer, which is a draft of the proposed course layout. By combining the tentative orientations of tees, fairways, and greens with the original topographic map, we could start to make calculations to estimate volumes of earth that must be moved to create the course. (Such cut-and-fill calculations are often considered the domain of civil engineering.) And once we have a tentative course layout, we can use a planimeter or map wheel to determine the length of each hole, which then provides us with the total course length (which is an important consideration for any golfer). Furthermore, based on a determination of the area of the holes, we can even begin to be able to calculate our needs for grass seed and fertilizer.

Overall, this process has involved a number of key steps. Several different kinds of spatial data were located, and then manipulated so that the important features in each were found at the same locations. Once these data were brought into a common geographic or spatial referencing system, it was possible to use them together, to develop a variety of types of derived information: the determination of potential corridors on the site, proposed locations for constructing facilities, and eventually, engineering estimates for earth moving equipment operators. As we will see, this is a very typical flow of data and information through a spatial data processing and analysis problem.

1.4 Applications

The number of data layers one needs to consider varies greatly from one application to another. Consider a more complex problem: deciding on the location of an airport. Some of the data layers or themes that a planner might require to site an airport include:

Administrative
- Land Ownership
- Government Jurisdiction
- Rights-of-Way
- Mining Claims
- Existing Land Use

Abiotic
- Surface Geology
- Subsurface Geology
- Surface Water
- Subsurface Water
- Flood Plains
- Archaeological Sites
- Elevation

Infrastructure
- Transportation Network
- Utility Corridors
- Zoning Restrictions

Biotic
- Endangered Species
- Vegetation Cover

Climatic
- Temperature
- Precipitation
- Fog
- Wind
- Photoperiod

Geographic information systems are used in a wide variety of settings. Landscape architects have embraced the concepts behind GISs for many years, analyzing site suitability and developing capabilities of planning for a specified use (McHarg, 1969). Civil engineers and architects involved in developing large sites have comparable interests and techniques, including considerations of environmental impacts such as noise perception and obscuring or changing views. Forestry professionals use this technology for site mapping and management, and for pest and disease monitoring. City planners are using geographic information systems to help automate tax assessment, emergency vehicle routing, and maintenance of transportation facilities and public lands.

Environmental managers and scientists use these systems for such applications as maintaining an inventory of rare and endangered species and their habitats, and monitoring hazardous waste sites. In addition to these kinds of applications, military planners add several more: gauging the ability of heavy vehicles to traverse different kinds of terrain, and determining which sites on military bases which are suitable for various kinds of training exercises. We discuss in more detail a few of these varied kinds of applications in Chapter 12.

1.5 Geographical Concepts

Before proceeding further, we will introduce a number of terms in common usage (based in part on the brief discussion in Van Roessel, 1987). We will return to some of these in more detail in chapter 3.

Spatial objects are delimited geographic areas, with a number of different kinds of associated attributes or characteristics. The golf course discussed above is a spatial object: it is a specific area on the ground, with many distinct characteristics (such as land use, tax rate, types of vegetation, number of parking spaces, etc.). On the golf course are a number of other spatial objects, such as the greens and fairways. A **point** is a spatial object with no area. The holes on our golf course represent points, even though they do in actuality cover a finite area. One of the key attributes of a point are its geodetic location, often represented as a pair of numbers (such as latitude-longitude, or northing-easting). There may be a range of data associated with a point, depending on the application. In our example, we may wish to record the number of the hole, as well as the date when a given hole on our golf course was placed on the green. The latter is useful so that we may remember to move the hole periodically to minimize wear on the green.

A **line** is a spatial object, made up of a connected sequence of points. Lines have no width, and thus, a specified location must be on one side of the line or the other, but never on the line itself. One important line in our example might indicate the out-of-bounds line between holes. Attributes we could attach to that line include the numbers of the holes that the line separates, and whether the line is indicated on the course by markers of a certain color. **Nodes** are special kinds of points, usually indicating the junction between lines or the ends of line segments.

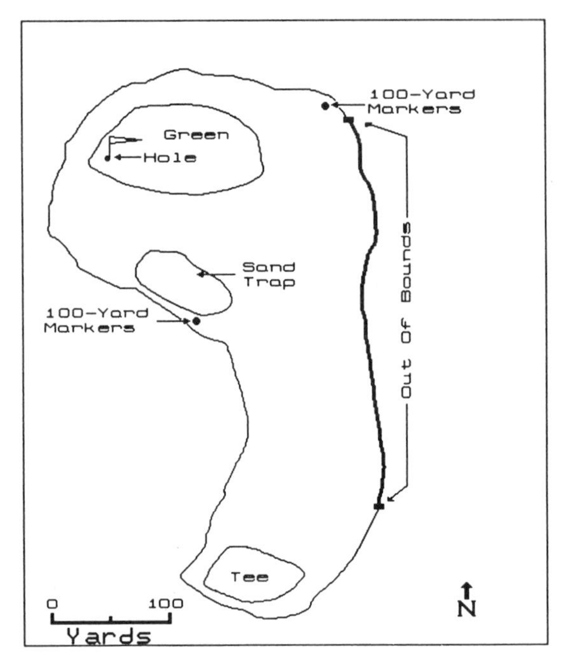

Figure 1. 3 Spatial Objects.

A **polygon** is a closed area. Simple polygons are undivided areas, while complex polygons are divided into areas of different characteristics. Since our example golf course hole has interior objects, such as the sand trap and the green, it is a complex polygon; since the sand trap is homogeneous (according to the available information in the figure), it is a simple polygon. Attached to the polygons on our golf course might be information about the length and area of each hole, and the kind and amount of seed and fertilizer used to maintain the fairways. **Chains** are special kinds of line segments, which correspond to a portion of the bounding edge of a polygon.

Figure 1.3 illustrates some of these different kinds of spatial objects, by focusing on one hole in our golf course. The boundary around the entire hole represents the boundary of a complex polygon. The location of the hole (or more specifically, its center) is a point. The 100-yard markers on either side of the fairway are certainly points, but since they form the ends of a line segment, we call them nodes. The portion of the out-of-bounds line that corresponds to the eastern edge of this hole would be considered a chain, since it corresponds to a portion of the polygon surrounding the entire hole.

We have already used the word **scale** in our discussions. By scale we mean the ratio of distances represented on a map or photograph to their true lengths on the Earth's surface. Scale values are normally written as dimensionless numbers, indicating that the measurements on the map and the earth are in the same units. A scale of 1:25000, pronounced *one to twenty five thousand*, indicates that one unit of distance on a map corresponds to 25,000 of the same units on the ground. Thus, one centimeter on the map refers to 25,000 centimeters (or 250 meters) on the Earth. This is exactly the same as one inch on the map corresponding to 25,000 inches (or approximately 2,080 feet) on the Earth. Note that scale always refers to linear horizontal distances, and not measurements of area or elevation.

The terms **small scale** and **large scale** are in common use. A simple example helps to illustrate the difference. Consider a field 100 meters on a side. On a map of 1:10000 scale, the field is drawn 1 centimeter on a side. On a map of 1:1,000,000 scale, the field is drawn 0.1 millimeter on a side. The field appears *larger* on the 1:10000 scale map; we call this a *large-scale map*. Conversely, the field appears *smaller* on the 1:1,000,000 scale map, and we call this a *small-scale map*. Said in another way, if we have a small area of the earth's surface on a page, we have a large-scale map; if we have a large area of the earth's surface on a page we have a small-scale map.

An important concept when working with spatial data is that of **resolution**. Most dictionaries define resolution in such terms of "distinguishing the individual parts of an object." For our purposes, however, we need a more specific definition. Tobler (1987) defines spatial resolution for geographic data as the content of the geometric domain divided by the number of observations, normalized by the spatial dimension. The **domain,** for two dimensional datasets like maps and photographs, is the area covered by the observations. Thus, for two-dimensional data, take the square root of the ratio to normalize the value. For example, if the area of the United States is approximately 6 million square kilometers, and there are 50 states, then the mean resolution element of a map portraying the states would be:

$$\begin{aligned}\text{mean resolution element} &= \sqrt{\text{area/number of observations}}\\ &= \sqrt{6x\ 10^6 km^2/50}\\ &= \textit{approximately}\ 346\ km.\end{aligned}$$

This gives us a way of examining some spatial data, and calculating a representative value for the spatial resolution of the dataset. If we increase the number of observations, the mean resolution element decreases in size. Consider a map of the United States that indicates each of the 3141 counties:

$$\text{square root of } (6 \text{ x } 10^6 \text{ km}^2/3141) = \sqrt{6x\ 10^6 km\ ^2/3141}$$
$$= \textit{approximately}\ \ 43\ km.$$

When we have more information, the mean resolution element gets smaller; we often call this a **higher resolution** dataset. Conversely, a lower resolution dataset will have fewer observations in an area, and thus, a larger mean resolution element. As we discuss in section 6.1.1, the size of the resolution element (sometimes abbreviated **resel**) is related to the size of the objects we can distinguish in a dataset.

For interested readers, a good discussion of other important concepts, including geometrical operations and relationships, may be found in Nagy and Wagle (1979).

1.6 The Four Ms

Our understanding of this planet has always been limited by our lack of information, as well as our lack of wisdom and knowledge. For things too small to see, we have developed microscopes that can image down to the molecular level. At the other end of the continuum, for things that are (in a very real sense) too large to see, we have geostationary satellites that can take an image of an entire hemisphere. Geographic information systems are a means of integrating spatial data acquired at different scales and times, and in different formats.

Basically, urban planners, scientists, resource managers, and others who use geographic information work in several main areas. They observe and **measure** environmental parameters. They develop **maps** which portray characteristics of the earth. They **monitor** changes in our surroundings in space and time. In addition, they **model** alternatives of actions and processes operating in the environment. These, then, are the four Ms: measurement, mapping, monitoring, and modeling (Figure 1.4). These key activities can be enhanced through the use of information systems technologies, and in particular, through the use of a GIS.

Geographic information systems have the potential for improving our understanding of the world around us. Yet these systems do *not* lessen the need for quality data, nor will these systems do the work for us. The work we

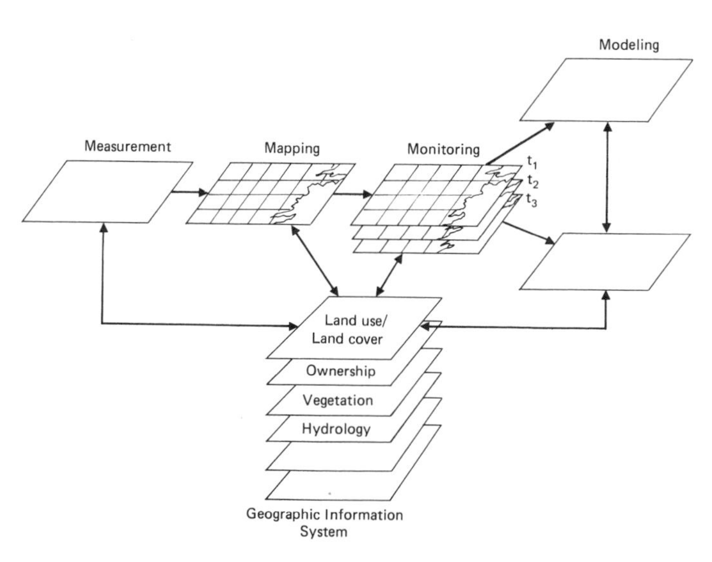

Figure 1.4 The four Ms. Measurement, mapping, monitoring, and modeling of environmental features and processes can be enhanced through the use of a geographic information system.

can do with a GIS is clearly dependent on the quality of data it contains. Thus, care must be taken to understand the potential sources and relative magnitudes of errors which may occur when gathering and processing spatial data. In addition, one must be cautious of the potential for misinterpretation of the information output from a GIS.

In interacting with a geographic information system, the user must not only understand the application, but also the characteristics of the tool and the system itself. Like all advanced technologies, the kinds of spatial data processing systems we will discuss must be employed wisely, to keep us from fooling ourselves. The following chapters discuss the wise use of geographic information systems.

Chapter 2

Background and History

Geographic information systems evolved as a means of assembling and analyzing diverse spatial data. Many systems have been developed, for land-use planning and natural-resource management at the urban, regional, state, and national levels of government agencies. Most systems rely on data from existing maps, or on data that can be mapped readily (Shelton and Estes, 1979).

The development of geographic information systems has its roots in at least two overlapping areas: an interest in managing the urban environment (particularly in terms of planning and renewal), and a concern for the balancing competing uses of environmental resources. Technology has played a critical role in addressing these concerns. If we look at John Naisbitt's 1984 work *Megatrends*, we can see why. *Megatrends* discusses new directions which are transforming our lives. In Naisbitt's words, none of the megatrends discussed "*is more subtle, yet more explosive than . . . the megashift from an industrial to an information society*." This information society had its beginnings in 1956 and 1957. Indeed, the advances in communications and computer technology that facilitated the widespread dissemination of the ideas and concepts contained in Rachel Carson's book *Silent Spring* also provided the foundations and requirements that necessitated the construction of automated geographic information systems.

Today, environmental scientists and resource managers have access to more data than ever. Naisbitt (1984) estimates that scientific information is doubling every five years. The key to coping with this information explosion is the employment of *systems* - systems that will take the data, analyze it, store it, and then present it in forms that are useful. These are the requirements of an information system.

2.1 The Cartographic Process

According to Robinson and Sale (1969), cartography is often described as a meeting place of science and art. This science/art is fundamentally directed at communicating information to a user and is central to an understanding of the strengths and weaknesses of geographic information systems technology. Much of the material contained in this book is directly related to essential elements of the cartographic process, which involves a body of theory and practice that is common to all maps.

Maps are both a very important form of input to a geographic information system, as well as common means to portray the results of an analysis from a GIS. Like a GIS, maps are concerned with two fundamental aspects of reality: locations, and attributes at locations. Location represents the position of a point in two-dimensional space. Attributes at a location are some measure of a qualitative or quantitative characteristic, such as land cover, ownership, or precipitation. From these fundamental properties a variety of topologic and metric properties of relationships may be identified, including distance, direction, connectivity and proximity. As Robinson et. al (1984) observe, *"a map is therefore a very powerful tool"*. Indeed, maps are powerful tools for communicating spatial relationships. Following Robinson et. al., maps:

- are typically reductions which are smaller than the areas they portray. As such, each map must have a defined relationship between what exists in the area being represented, and the mapped representation. This relationship is of primary importance. Scale sets limits on both the type and manner of information that can be portrayed on a map.

- involve transformations. Often in mapping, we are faced with a need to transform a surface which is not flat (such as a portion of the earth's surface). In order to represent such a surface on a flat plane, map projections are employed (see section 6.6.1). Choice of a particular projection has an impact on how a given map may be used. Plane coordinate grids are often used on maps as systems of reference.

- are abstractions of reality. Maps are the cartographer's representation of an area, and as such, display the data that the cartographer has selected for a specific use. Thus, the information portrayed on a map has been classified and simplified to improve the user's ability to work with the map.

- contain symbols which represent elements of reality. Few map symbols have universally accepted meanings, but some maps use a standardized set of symbols.

- portray data using a variety of marks, including lines, dots, tones, colors, textures, and patterns.

In addition to these basic characteristics of maps, the user of maps and other products of a geographic information system should understand the errors which may affect them. The sources of errors fall into three categories (Burrough, 1986): obvious sources, those resulting from natural variation and original measurements, and those arising through processing.

Obvious sources:

The source data may be too old to be of value.

The areal coverage of a given data type, within a given time frame, may not be complete.

The scale of the map may restrict the type, quantity, and quality of the data which may be presented.

The number of observations within the target area may not be sufficient to be able to determine the spatial patterns in the objects of interest.

Practical matters such as the time, funds, and staff which are available may not permit us to produce a product of the required characteristics.

Natural variation and the original measurements:

Positional accuracy of the source data may not be sufficient, due to problems in the field data itself, instrument errors, and lack of rigor in the compilation process.

Attribute errors also may come from a variety of sources, including both mis-identification and compilation problems.

Processing:

Numerical errors may include round-off or dynamic-range errors in arithmetic computations.

Errors in logic may cause us to manipulate the data incorrectly, thus leading us to fool ourselves. Common problems in this area are associated with classification and generalization.

The above lists focus on errors in a single map sheet, or a single GIS data layer. When working with many layers at once, the separate layers may not be completely compatible in terms of scale and accuracy, thus complicating the task of creating an accurate, final, analytic product. This, as well as other problems, such as developing efficient and cost-effective means for verifying the accuracy of map and GIS products, are active research areas.

2.2 Early History

Cartography is defined in the *Multilingual Dictionary of Technical Terms in Cartography* (Meynen, 1973) as "*the art, science and technology of making maps together with their study as scientific documents and works of art.*" Map-making *per se* can be traced back to the ancient cultures of Mesopotamia and Egypt. The earliest known map, a regional map imprinted in a clay tablet, dates from about 2500 B.C. Yet people must have been making maps much earlier than that. Simple arrangements of sticks or pebbles were probably used to illustrate geographic relationships long before clay tablets or papyrus came into vogue. A mound of dirt, a few pebbles, and a small furrow made with a stick, could have illustrated important game trails or berry-picking locations, and could thus have been the first analog geographic information system.

Parent and Church (1988) state that the origins of more sophisticated geographic information systems go back to early developments in cartography. They reference the mid-eighteenth-century production of the first accurate base maps as an important point in GIS development. As Parent and Church point out, until the development of high-quality base maps, the accurate graphic depiction of spatial attributes was not possible. These developments were followed by a rapid expansion of the use of thematic mapping. The idea of recording various layers of spatial data on a series of similar base maps was an established cartographic convention by the time of the American Revolutionary War (Harley et al., 1978). For example, a map by French military leader and cartographer Louis Alexandre Berthier (1753-1815) contained hinged overlays showing troop movements during the 1781 Siege of Yorktown (Rice and Browns, 1974).

In the early part of the nineteenth century, advances in both the physical and social sciences provided geographers with important intellectual tools for the analysis of spatial data. Such fields as statistical analysis, number theory, and advanced mathematics flourished. The first geologic maps of London and Paris appeared. The work of the distinguished German geographer Alexander Freiherr von Humbolt (1769-1859) became influential. The British census of 1825 produced a tremendous amount of data to be analyzed, and the science of demography soon evolved.

Church and Parent (1988) state that by 1835, technology (in particular, advanced cartographic techniques), social science, and social thought (specifically, concepts of environmental responsibility) had progressed to support new and improved levels of thematic mapping. However, it was the economic changes of the industrial revolution, according to Church and Parent, that provided the main catalyst for the early evolution of geographic information systems in this time period. The explosion in manufacturing, with the attendant demand for raw materials and labor, created the need for a new, extensive infrastructure, both social and industrial. Indicative of all these changes is a transportation study, completed in 1837, that first brought together technical, social, and scientific advances related to spatial data analysis, into a coherent whole. The *Atlas to Accompany the Second Report of the Irish Railway Commissioners*, appearing in 1838, consisted of a series of maps with a uniform base, depicting population, traffic flow, geology, and topography.

As this example indicates, cartographers realized some 150 years ago that a single map may not contain all the data required to satisfy a given information need. Indeed, the data may not exist in map form at all, but in graphs, text, or statistical tables. As researchers and resource managers began to ask more and more complex questions about their environment, their need for improved methods of processing spatially distributed data increased. This need lead to the beginnings of automated geographic data processing in the late 1800's.

Streich (1986) states that American statistician Herman Hollerith (1860-1929) was the father of automated geoprocessing. Hollerith adapted punched-card techniques, which had been used in France to program looms, to help process the information collected in the 1890 United States Census. Hollerith conceived the idea of punching raw demographic data onto cards and using machines to sort and collate this data. Streich goes on to state that "... the move to 'electro-mechanical' data-processing technology for census tabulation characterizes a fundamental tenet of geoprocessing -- that a need exists to rapidly, accurately, and cost-effectively collect, analyze, and distribute spatially disposed information."

In 1936, Charles Colby's presidential address to the Association of American Geographers (Colby, 1936) laid out research challenges in geography. Among these challenges particular emphasis was placed on the development of quantitative approaches to map-based problems. In doing so, Colby set the stage for the modern era of geographic information systems.

2.3 The Modern Era

From these early beginnings, advances in computing, cartography, and photogrammetry laid the technological foundations for the automated geographic information systems that began to appear in the 1960s. The conceptual framework within which early geographic information systems were implemented involved individuals in many disciplines. Researchers and resource managers in diverse areas realized that there was a need for integrating data from a variety of sources, to manipulate the sets of data to analyze them, and then to be able to provide information for a resource planning and management decision process.

Three important factors helped lead to the creation of digital geographic information systems in the 1960's:

- refinements in cartographic technique,
- rapid developments in digital computer systems, and
- the quantitative revolution in spatial analysis.

These developments were very important in that they helped to provide analytic tools as well as stimulation to researchers and professionals in a variety of applications. In addition to these, we must not forget the advances in geographic thought that helped to bring about the modern GIS. Chrisman (1988) points out that the "*cult of novelty and high technology*" can blind us as to various disciplines' contributions to GIS development. As a geographer, he points to Sauer's early work in the upper peninsula of Michigan, as well as to the land-use research of Whillesey, Finch, and others that foreshadowed our current paradigm of stacking layers of thematic information. The roots of such map overlay may be traced back at least 100 years in the field of landscape architecture (Steinitz et al., 1976).

In 1969, Ian McHarg's *Design with Nature* was published. This work formalized the concept of land suitability/capability analysis (*SCA*). SCA is a technique in which data concerning land use in a locale being studied is entered into an analog or digital GIS. SCA programs are used to combine and compare data types via a deterministic model, in order to produce a general plan map. If the model is carefully implemented, and suitable data is available, this map should be consistent with existing land-use classes and the constraints that are imposed by both natural and cultural features. *Design with Nature* is a seminal work, influencing the use of overlays of spatially referenced data layers in the resource planning and management decision process. McHarg's efforts in SCA have been followed by many articles in this area.

Resource management concerns spurred development of spatial data-processing systems in the U.S. government during the past two decades. A system called *STORET* was developed by the Public Health Service for storage of spatial information about water quality (Green, 1964). Another, called *MIADS*, was developed by the U.S. Forest Service for the analysis of recreation alternatives and hydrology (Amidon, 1964). The Census Bureau was also heavily involved in geocoding and automated spatial data processing at this time.

In the university community at this time, Harvard's Laboratory for Computer Graphics developed and made available a series of automated mapping and analysis programs. The University of Washington at Seattle also made important contributions, particularly in the areas of transportation analysis and urban planning and renewal (Gaits, 1969). Urban planning applications blossomed with the development of these kinds of tools; by 1968 thirty-five urban and regional planning agencies in the United States were using automated systems (Systems Development Corp., 1968).

The first system in the modern era to be generally acknowledged as a GIS was the Canada Geographic Information System or *CGIS* (Peuquet, 1977). Roger Tomlinson (1982), involved in the design and development of the system, states that CGIS was designed specifically for the Agricultural Rehabilitation and Development Agency Program within the Canadian government. The main purpose of CGIS was to analyze Canadian Land Inventory data, which was being collected to find marginal lands. Therefore, in the broadest sense, the first geographic information system was developed to help with an environmental problem: rehabilitation and development of Canada's agricultural lands.

The Canadian Geographic Information System was implemented in 1964 (Deuker, 1979). This was one year after the first conference on Urban Planning Information Systems and Programs, a conference which led to the establishment of the Urban and Regional Information Systems Association. The New York Landuse and Natural Resources Information System was implemented in 1967, and the Minnesota Land Management Information System in 1969. In these early years, the costs and technical difficulties of implementing a GIS prevented all but national- and state-goverment agencies from developing these systems.

In 1977, a report issued by the United States Department of the Interior's Fish and Wildlife Service compares the selected operational capabilities of 54 GISs (USFWS, 1977). This survey, which is representative of several others conducted in the late 1970s, provides information on the hardware environment, programming language, documentation, and characteristics of the systems. This survey lists many GISs developed by federal and state agencies, as well as universities. However, it contained information on only a few commercial GISs. Even today, few commercial firms offer fully integrated geographic information systems. Streich (1986) estimates that there are ten commercial firms offering GISs on the open market.

Why is it that there are so few commercial firms offering geographic information systems, even today? There is no simple answer. A GIS is a complex hardware and software system, and requires considerable expertise in a variety of geographic, computer science and systems engineering areas. Nevertheless, some GISs are being developed in the private sector. These commercially available systems provide a wide range of well-integrated GIS capabilities. As technology and scientific understanding improve, the development of geoprocessing systems becomes more and more open to commercial firms. Instead of being a large-user in-house activity, the development of geoprocessing systems will likely be taken over by commercial firms and made available to a variety of hardware environments, discipline interests, and goals.

In addition to the beginnings of commercial GIS development, the 1970s also saw significant developments in image processing and remote sensing systems, which often had some GIS functions. Such firms as the Environmental Systems Research Institute in California began operation. Image processing systems with some GIS elements were developed at the Jet Propulsion Laboratory and at the Purdue University Laboratory for Applications of Remote Sensing. These latter systems incorporated GIS capabilities as the remote sensing community quickly realized that ancillary GIS data could play an important role in improving the accuracy of the interpretation of remotely sensed data.

Developments in remote sensing technology and applications during this decade spurred practical and theoretical work in the areas of geometrical corrections and registration. The coupling of map and image data also drove work in raster-vector data format conversion (see Chapter 6). Today there are both commercial and public-domain image processing systems that possess sophisticated GIS capabilities; however, we still believe that a great deal more work needs to be done in terms of effectively integrating remotely sensed data into traditionally vector-based geographic information systems.

An interesting footnote to this brief history of geographic information systems comes from 1982 conference, "Computer Assisted Cartography and Geographic Information Processing: Hope and Realism," held at the University of Calgary. In a session led by Roger Tomlinson, several individuals were critical of the operational status of CGIS. The criticisms revolved around the question of whether the system was meeting the needs of the users. Tomlinson responded that the system was originally designed for a class of users who wished to analyze land inventory data and find marginal farms. This class of users had effectively disappeared, and there were now over 100 other users and a nine-month backlog of work. He concluded that this tremendous number of users and long backlog indicated that the system was successful. In response to Tomlinson's argument, it was noted that for many applications, a nine-month backlog was intolerable; when users cannot get their work done on the system, something is indeed not right.

The message we derive from this story is that problems will certainly arise as geographic information systems designed and implemented for a specific class of problems are used for other purposes. Users and system managers must guard against inappropriate use of systems and must establish priorities and long-range upgrade and migration policies to meet the needs of changing user communities and changing data.

In summary, the development of geographic information systems, in terms of both the underlying concepts and the technology, has drawn on the talent and experience of many researchers and investigators. It has grown out of concerns about the state of the physical and cultural environment, and it has been advanced by efforts in both the public and private sectors. Many early systems were developed to solve relatively narrow, specific kinds of problems. The past twenty years have seen an explosion in the technological base for these systems, particularly in the areas of data processing and remote sensing systems. The 1980s have seen continued growth in GIS applications, significant system refinements, and a modest expansion of the commercial availability and applicability of geographic information systems.

While many operational systems may be limited in terms of the geographic area, the number of data types, and the modeling and analytic capabilities they can provide, they can perform many operations that only 25 years ago were considered unfeasible. One recent trend in the evolution of GIS technology is the inclusion of artificial intelligence into GIS design and operation. This topic was the subject of a workshop at a recent international symposium in Zurich, Switzerland (Smith, 1984). We will examine some of these far-reaching discussions in section 12.8. Let us now consider some of the generic components and functions of geographic information systems.

Chapter 3

The Essential Elements of a Geographic Information System: An Overview

As we said in the first chapter, an information system is fundamentally an end-to-end system, which deals with the flow of data and information from its primary sources to the derived information and its ultimate uses. **Geographic** information systems are designed to handle information regarding spatial locations. In this chapter, we will introduce the essential functional components of a GIS, and will discuss some key concepts in geography and geographic data processing.

3.1 GIS Functional Elements

There are five essential elements that a GIS must contain (Figure 3.1; based on the discussion in Knapp, 1978): data acquisition, preprocessing, data management, manipulation and analysis, and product generation. For any given application of a geographic information system, it is important to view these elements as a continuing process. We will introduce each of the elements in this chapter, and will examine each in greater detail later in this text. As a guiding principle, the analyst should develop an end-to-end model of the task at hand. Even when the precise details of the steps to be taken may depend on the results of intermediate calculations and analyses, an explicit outline of the process, like a working hypothesis in a scientific experiment, can be very valuable.

Data acquisition is the process of identifying and gathering the data required for your application. This typically involves a number of procedures. One procedure might be to gather new data by preparing large-scale maps of natural vegetation from field observations, or by contracting for aerial

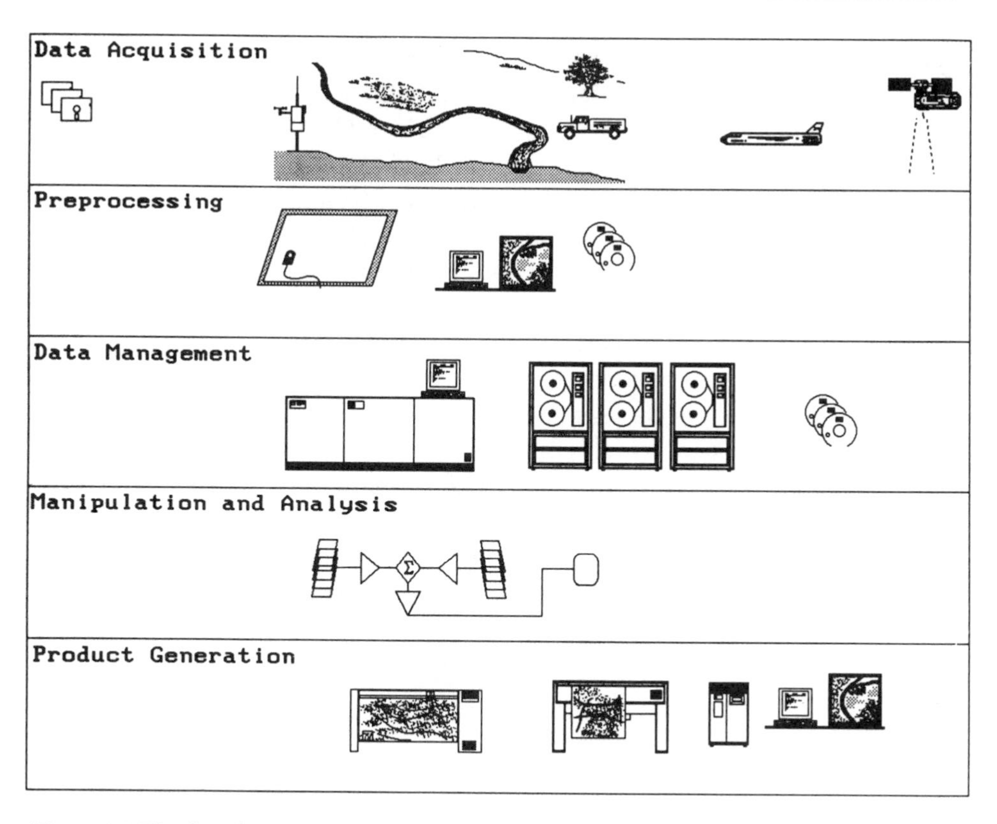

Figure 3.1 The five elements of a GIS.

photography. Other kinds of surveys may be required to determine, for example, consumer satisfaction and preferences in different parts of a city to help locate new business offices. Other procedures for data acquisition may include locating and acquiring existing data, such as maps, aerial and ground photography, surveys of many kinds, and documents, from archives and repositories.

One must *never* underestimate the costs (in time as well as money) of the data-acquisition phase. A GIS is of no use to anyone until the relevant data have been identified and located. Furthermore, the accuracy of the decisions reached through spatial analysis is limited by the accuracy and precision of the underlying datasets. We often know too little about the underlying quality of many kinds of spatial data. At times, however, we may be forced to use maps and other datasets whose underlying quality is unknown. And without spending some effort ensuring that various datasets are not only relevant but also reliable, we run the risk of fooling ourselves.

Preprocessing involves manipulating the data in several ways so that it may be entered it into the GIS. Two of the principal tasks of preprocessing include data format conversion and identifying the locations of objects in the original data in a systematic way. Converting the format of the original data often involves extracting information from maps, photographs, and printed records (such as demographic reports) and then recording this information in a computer database. This process is a time-consuming and costly effort for many organizations. This is particularly (and sometimes painfully) true when one calculates the costs of converting large volumes of data based on paper maps and transparent overlays, to an automated GIS based on computerized datasets. We will discuss aspects of the this process in section 6.1.

A second key task of the preprocessing phase is to establish a consistent system for recording and specifying the locations of objects in the datasets. When this task is completed, it is possible to determine the characteristics of any specified location in terms of the contents of any data layer in the system. During these processes, it is very important to establish specific quality control criteria for monitoring the operations during the preprocessing phase, so that the databases can be of maximum value to the user.

Data-management functions govern the creation of, and access to, the database itself. These functions provide consistent methods for data entry, update, deletion, and retrieval. Modern database management systems isolate the users from the details of data storage, such as the particular data organization on a mass storage medium. When the operations of data management are executed well, the users usually do not notice. When they are done poorly, everyone notices: the system is slow, cumbersome to use, and easy to disrupt. Under these latter circumstances, the smallest human and machine errors create large problems for both the users and the system operators. Data-management concerns include issues of security. Procedures must be in place to provide different users with different kinds of access to the system and its database. For example, database update may be permitted only after a control authority has verified that the change is both appropriate and correct.

Manipulation and analysis are often the focus of attention for a user of the system. Many users believe, incorrectly, that this module is all that constitutes a geographic information system. In this portion of the system are the analytic operators that work with the database contents to derive new information. For example, we might specify a region of interest and request that the average slope of the area be calculated, based on the contours of elevation that have already been stored in the GIS database. Since no single system can encompass the complete range of analytic capabilities a user can imagine, we must have specific facilities to be able to move data and information between systems. For example, we may need to move data from our GIS to an external system where a particular numerical model is available, and then transport the derived results back into the spatial database inside the GIS. This kind of modularity, where other data processing and analysis systems can be linked to a GIS, is very valuable in many circumstances, and permits the system to be easily extended over time by pairing it with other analytic tools. When one speaks of *geoprocessing*, one is often focused on the manipulation and analysis components of a GIS.

Product generation is the phase where final outputs from the GIS are created. These output products might include statistical reports (such as a table listing the average population densities for each county in California, or a report indicating landowners who are delinquent in their property taxes), maps (for example, a presentation of the property boundaries of plots within a township that are owned by public agencies, or a map of a subdivision indicating where construction workers must be careful when digging due to the presence of underground pipes and cables), and graphics of various kinds (such as a set of bar charts that compare the acreage of different crop types in an area). Some of these products are **soft copy** images: these are transient images on television-like computer displays. Others, which are durable since they are printed on paper and film, are called **hard copy**. Increasingly, output products include computer-compatible materials: tapes and disks in standard formats for storage in an archive or for transmission to another system. The capability of taking the output of an analytic process, and placing it back into the geographic database for future analysis, is extremely important.

These essential components of a *geographic* information system are the same as those of any other information system. Let us compare this sequence of functional elements to a more conventional information system problem. Consider the steps that are taken in an automated system to manage employee records for a business. Information about the individuals must be gathered together, perhaps via a questionnaire and interview when the individual is hired. This is clearly the **data acquisition** phase. Then, because some of the information is inevitably expressed by different people in different ways (for example, some people will list their education as "through grade 12", while others will say "through high school"), the data must be put into a consistent vocabulary and format. Only after this **preprocessing** phase can the data be entered into the computer in a consistent form. Validation of the data entered into the system is a fundamental part of the preprocessing phase, to insure the accuracy of the resulting database.

Once the data have been converted into a consistent form and put in the computer database, we have accomplished a large fraction of the end-to-end task and often expended a large fraction of the end-to-end costs. **Data management** functions permit us to update the information when necessary (for example, when an employee completes an advanced degree), and to retrieve only the relevant information when required (as in a summary report of salaries for a particular division of the company). Various kinds of **analytical operations** can be run -- perhaps using employee addresses to find out which employees live close to one another in an effort to encourage car pooling. Finally, we need to be able to develop statistical reports, graphics of many kinds, and other **output products**, such as documentation for management reviews of salary levels. These steps exactly parallel the five GIS components we will discuss in detail.

3.2 Data in a GIS

It is important to understand the different *kinds* of variables that can be stored in any information system. **Nominal** variables are those which are described by name, with no specific order. Categories of land use (such as parks, wilderness areas, residential districts, and central business districts) and trees (such as *Eucalyptus calophylla*, *Pinus coulteri*, and *Quercus agrifolia*) are different kinds of nominal variables. These are common in many kinds of thematic maps. **Ordinal** variables are lists of discrete classes, but with an inherent order. Classes of streams (first order, second order, and so forth; referring to the number of tributaries which contribute to the stream) or levels of education (primary, secondary, college, post-graduate) are ordinal variables since the discrete classes have a natural sequence. **Interval** variables have a natural sequence, but in addition, the *distances* between the values have meaning. Temperature measured in degrees Celsius is an interval variable, since the distance between 10°C and 20°C is the same as the distance between 20°C and 30°C. Finally, **ratio** variables have the same characteristic as interval variables, but in addition, they have a natural zero or starting point. Since degrees Celsius is a measurement with an arbitrary zero point, the freezing point of pure water, it fails the latter test. Degrees Kelvin, since it is based on an absolute standard, is ratio variable. Per capita income, the fraction of the weight of a soil sample that passes through a specified sieve, and rainfall per month are common ratio variables.

In addition to these 4 kinds of data, there are two different classes of data found in most geographic information systems. Consider a simple object in space: a water well. From the point of view of a GIS, the primitive but essential piece of information to record about this water well is its location on the Earth -- a data value pair such as longitude and latitude, thus storing the simplest kind of spatial data. However, there may be a wide range of additional information which is required for many applications. This might include the depth of the well, the volume of water produced over a given period of time, dates of pump tests, and temporal sequences of measurements of dissolved and particulate matter in the water from the well. This second set of **non-spatial** or **attribute** data, which is logically connected to the spatial data, must not be forgotten. In many geographic information systems, there are tools to both store and manipulate the non-spatial data along with the spatial data. In some applications, as we will see, the volume of non-spatial data may actually be larger than the volume of the spatial data, and the logical connections between the spatial and non-spatial information may be very important.

A recent issue of *The American Cartographer* (January, 1988), the journal of the American Congress on Surveying and Mapping, proposes a standard for digital cartographic data. This standard is based on entities in the real world, and a mechanism to represent these entities in terms of objects in a database. Within this proposal is a set of definitions of spatial objects, which we now paraphrase to explain more of the vocabulary of geographic information systems. This brief discussion also expands on the comments in Chapter 1 about different kinds of spatial objects. One may divide the different kinds of spatial objects into three classes, based on spatial dimensions of the objects.

A 0-dimensional object is a point that specifies a geometric location. From a mathematician's perspective, a point is a primitive location with no areal extent. Points are used in a number of ways in both computer graphic and digital cartographic data, as well as in a geographic information system. They are commonly used to indicate features themselves, such as the exact center of the water well mentioned above, the end of a street, or the corner of a lot in a subdivision. Points are also used as a reserved position for a label (such as a place name) or a symbol (such as an airport or benchmark) on a map, or to carry information for the surrounding region (such as who owns the region, or the color to be used when the region is displayed). Points are also used to define more complex spatial objects, such as lines and areas.

The simplest 1-dimensional object is a straight line between two points. More complex forms of lines include connected sets of straight lines (determined by the sequence of points at which the path changes direction), curves which are based on mathematical functions, and lines whose direction is specified. Particular sets of mathematical functions are used to define curves in some disciplines, as in the functional definition of the curve of a street used by a civil engineer. One advantage of a directed line segment is that we have a way to distinguish which end is the beginning of the line, and which end is the end. This may be particularly valuable in circumstances as diverse as the analysis of flow in pipes (perhaps indicating source and destination for flow in a potable water supply system) or models of population flow between countries. When the line segments carry information about direction, we are also able to distinguish the regions on the left and right sides of the line. As we shall see later, this can be very useful in a number of applications.

Finally, 2-dimensional objects are areas, which also come in many forms. In a particular application, we may refer to a bounded area, or focus on just the boundary, or just the region within the boundary. The description of the area itself is normally based on the geometry of the bounding line segments. The area may be either homogeneous or divided internally, as discussed in Chapter 1. A distinction is often made between sets of two-dimensional bounded regions, and true three-dimensional surfaces. In some applications, an analysis based on a two-dimensional planimetric representation of the Earth may be completely sufficient. We focus on these kinds of applications in this introductory text.

The details of the connections between spatial objects, such as the information about which areas bound a line segment, is called **topology**. One of the distinguishing features of some geographic information system databases is that they have explicit mechanisms to store topology, as we shall see in Chapter 4.

Cowen (1987) discusses a geographic information system from several different points of view. The *database approach* stresses the ability of the underlying data structures to contain complex geographical data. The descriptions of spatial objects in the previous several paragraphs take this view. In Chapter 4 we examine a number of common alternatives to storing spatial data. The *process-oriented approach* focuses on the sequence of system elements used by an analyst when running an application -- the five components we discussed at the beginning of this chapter follow this view. Chapters 5 through 9 in this text represent such an approach. An *application-oriented approach* defines a GIS based on the kinds of information manipulated by the system and the utility of the derived information produced by the system. Chapter 12 presents a number of uses of these spatial data processing systems, and clearly emphasizes this view. A natural resources inventory system is an easily understood example of this approach. Finally, a *toolbox approach* emphasizes the software components and algorithms that should be contained in a GIS. We develop a number of details from this point of view of a GIS in Chapters 6 and 8. Each of these different points of view of a geographic information system is useful; we recommend that the reader consider the differences between them during the following discussions.

Chapter 4

Data Structures

There are a number of different ways to organize the data inside any information system. The choice of a particular **spatial data structure** is one of the important early decisions in designing a geographic information system. While very few of us will ever design a GIS from start to finish, a knowledge of data structures is valuable from several points of view other than system design. Fundamentally, users must be aware of the characteristics of several different structures, since several different standard forms are commonly used, and the choice of a data structure can affect both data storage volume and processing efficiency. From another point of view, when we are collecting our own data, we must make a choice of data structure for storage. Also, in an operational or research environment, it is often necessary to convert datasets between several different data structures, either to work with several kinds of data at the same time, or to import an unusual dataset into an existing system. It is very important to be able to understand how these conversions affect the underlying information itself.

As we stated in the introduction, each different type kind of spatial data or **theme** in a GIS is referred to as a data layer or data plane. In each of these data layers there are three primitive types of geometrical entities to encode (after Peucker and Chrisman, 1975): points, lines, and polygons or planes. Some authors make a distinction between the representation of a truly three-dimensional surface, such as elevation datasets, and a representation of space in two dimensions, such as legal boundaries of land ownership on a flat map. In this chapter, we will focus on the latter.

The essential function of the spatial data we store and manipulate is to subdivide the Earth's surface into meaningful entities or objects that can be characterized. In this way, the contents of a spatial database is a *model* of the Earth. **Points**, such as the locations of oil and water wells, and **lines**, such as the centerlines of roadways or streams, are key elements of this breakdown into component parts. When we consider bounded regions, such as the borders of a subdivision or the edges of a reservoir, we often focus on the boundary **lines**, and call the enclosed regions **polygons**. These polygonal regions are not necessarily defined in the precise terms of geometry, where a polygon is ordinarily a planar figure bounded by a series of straight line segments. In spatial data processing, common usage relaxes the requirement that the bounding line segments be straight; we use the term polygon even when the boundaries are curved. We note, however, that not all GISs can work directly with curves as such, but more often permit a single line to have interior digitized points in addition to the end points. Many sophisticated applications have been developed around networks of lines, such as the network developed by the arteries of a transportation or communications system, or a variety of piping systems such as a sanitary sewer or a pressurized gas delivery system.

The above discussion concentrates on the geometry of the data. Equally important is the non-spatial or **attribute** data, which in some systems requires a greater amount of data storage. For a simple spatial object like a water well, the essential spatial information is the geodetic location of the well. The attribute data can include wide range of ancillary information about that well, including its depth, date of drilling, production volume, ownership, and so forth. Many geographic information systems have specialized capabilities for storing and manipulating the attribute data in addition to the spatial information.

4.1 Raster Data Structures

One of the simplest data structures is a **raster** or cellular organization of spatial data. In a raster structure, a value for the parameter of interest -- elevation in meters above datum, land use class from a specified list, plant biomass in grams per square meter, and so forth -- is developed for every cell in a (frequently regular) array over space. For example, in Figure 4.1, elevation in meters above mean sea level has been recorded at locations on a regular grid. The original data is from a topographic map, from which we have extracted the contour lines. The raster array of elevations is derived from these contour lines using procedures discussed in section 6.7. This kind of data structure is intuitive; we might imagine a survey team determining elevations at regular distances along lines of constant latitude.

4.1.1 Simple Raster Arrays

The horizontal dimension of the simplest raster, along the **rows** of the array, is often oriented parallel to the east-west direction for convenience. Following the conventional practice in image processing, raster elements in this direction along the rows of the array are sometimes called samples, and numbered from the left (or west) margin. Positions in the vertical direction, aligning with the **columns** of the array, are often numbered starting from the top (or northern) boundary. This numbering scheme comes from the computer graphics field, in which displays are often painted on the computer screen or printer from the top down. Thus, the origin of the raster is frequently the upper left corner. This location is considered position (1,1) in some systems of notation, and position (0,0) in others - please be aware of the difference!

Note that this referencing system for cells in a raster is different from more traditional georeferencing systems such as latitude-longitude in which one specific point on the Earth's surface (such as the point where the prime meridian crosses the equator) is the origin. It is also different from the Universal Transverse Mercator system, where (in the northern hemisphere) the origin of the coordinate system is in the lower left corner, which is similar to a conventional cartesian system. Often, the distances between cells in the raster are constant in both the row and column directions; in other words, the cells in the raster are square. In this case, it is natural to store the data on a computer in a two-dimensional array.

While a simple rectangular raster structure is a very popular approach, there are at least two limitations. First, for a raster structure at a particular scale, there is a finite limit to our ability to specify location. We are in either one cell or another - there is nothing in between. This is true because the line separating adjacent cells is considered to be infinitely narrow. This is the case for any raster data structure, including the non-rectangular forms we'll examine in a moment.

Second, adjoining cells may not be evenly spaced, depending on how we define the word *adjoining*. Consider the center cell in Figure 4.2a. The cells above and below are 1 unit of distance from the center cell, while the cells on the diagonal are approximately 1.41 (the square root of 2) units of distance from the center. For searches through the data, if we include only cells above

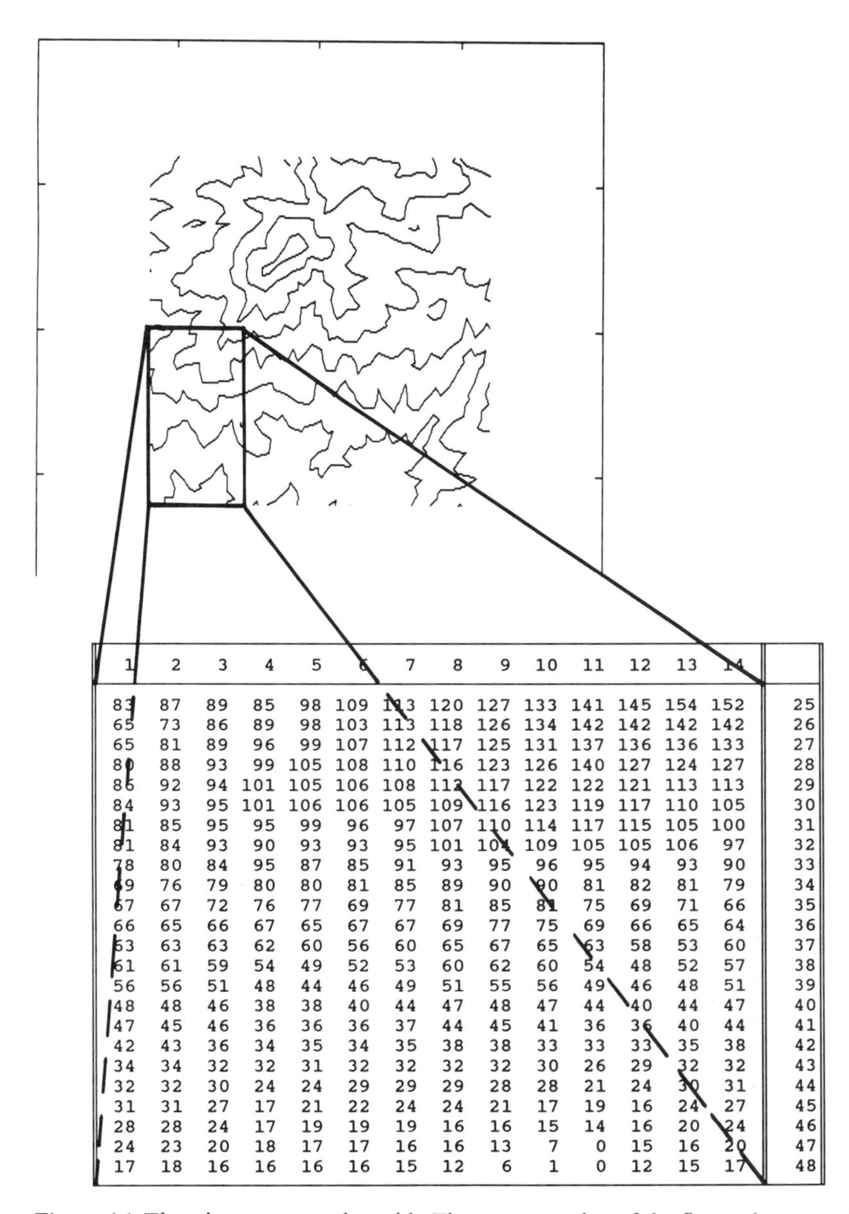

1	2	3	4	5	6	7	8	9	10	11	12	13	14	
83	87	89	85	98	109	113	120	127	133	141	145	154	152	25
65	73	86	89	98	103	113	118	126	134	142	142	142	142	26
65	81	89	96	99	107	112	117	125	131	137	136	136	133	27
80	88	93	99	105	108	110	116	123	126	140	127	124	127	28
86	92	94	101	105	106	108	112	117	122	122	121	113	113	29
84	93	95	101	106	106	105	109	116	123	119	117	110	105	30
81	85	95	95	99	96	97	107	110	114	117	115	105	100	31
81	84	93	90	93	93	95	101	104	109	105	105	106	97	32
78	80	84	95	87	85	91	93	95	96	95	94	93	90	33
69	76	79	80	80	81	85	89	90	90	81	82	81	79	34
67	67	72	76	77	69	77	81	85	81	75	69	71	66	35
66	65	66	67	65	67	67	69	77	75	69	66	65	64	36
63	63	63	62	60	56	60	65	67	65	63	58	53	60	37
61	61	59	54	49	52	53	60	62	60	54	48	52	57	38
56	56	51	48	44	46	49	51	55	56	49	46	48	51	39
48	48	46	38	38	40	44	47	48	47	44	40	44	47	40
47	45	46	36	36	36	37	44	45	41	36	36	40	44	41
42	43	36	34	35	34	35	38	38	33	33	33	35	38	42
34	34	32	32	31	32	32	32	32	30	26	29	32	32	43
32	32	30	24	24	29	29	29	28	28	21	24	30	31	44
31	31	27	17	21	22	24	24	21	17	19	16	24	27	45
28	28	24	17	19	19	19	16	16	15	14	16	20	24	46
24	23	20	18	17	17	16	16	13	7	0	15	16	20	47
17	18	16	16	16	16	15	12	6	1	0	12	15	17	48

Figure 4.1 Elevations on a regular grid. The upper portion of the figure shows a set of contour lines from a map, in which locations along a given line are the same elevation above sea level. The array of numbers in the lower portion of the figure represent the estimated elevation at the center of a set of square raster cells.

and below a cell of interest, we are working in a **4-connected** neighborhood (Figure 4.2a), and all cells are equidistant from their neighbors. Thus, neighboring cells share an edge. If we include elements on the diagonal, we are working in an **8-connected** neighborhood (Figure 4.2b), and now cells are not evenly spaced. In this latter case, some cells in the neighborhood share an edge, while others share only a vertex. Since all the cells in these two examples have neighbors of the same size and shape, we say that we have **spatial neighborhood similarity** (Tobler, 1979). An alternative is that each cell in the neighborhood has a different size and organization. While this is certainly reasonable for many kinds of geographic phenomena (urban neighborhoods are of different sizes and shapes, for example), it is not a common data model in a raster geographic information system.

Whether it is more appropriate to base an analysis on 4-connected or 8-connected space must be determined by the characteristics of the data and the objectives of the exercise. In many metropolitan areas of recent design, the streets are laid out in a rectangular network. If a raster data layer is used to represent this transportation network, a 4-connected analysis of travel distances is appropriate, since diagonal motion is not ordinarily possible. In contrast, an 8-connected analysis of travel might be appropriate for a raster database of the weight of cotton harvested per hectare, where the objects traveling through space are airborne insect pests that do not respect the orientation of the local roads.

It isn't necessary that the size of a raster be the same in the row and column directions. But for reasons of simplicity and symmetry, applications based on the use of rectangular raster cells are rare. It is also important to recognize that there are two different theoretical interpretations of the value stored in a cell of our sample raster of elevations. A cell's value might

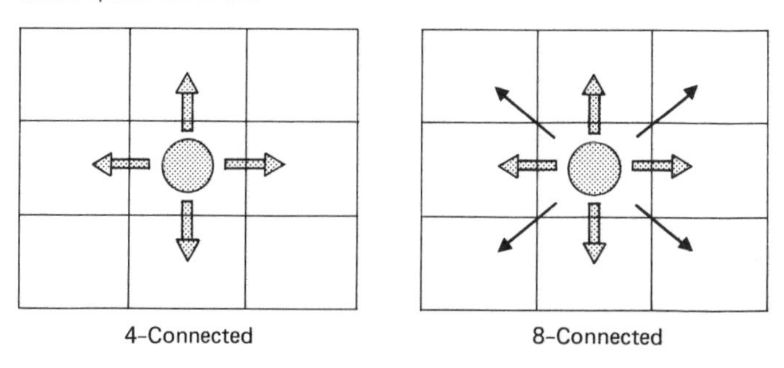

Figure 4.2 Spatial neighborhoods. Cells which share a side are, on average, closer to each other than those which share only a vertex.

represent an elevation measured at the center of the cell, or the value might represent the average elevation of the entire cell. It is likely that neither interpretation is truly correct, although we typically operate as if the latter is our underlying model of truth. For practical purposes, the distinction rarely matters if we've chosen an appropriately small cell size.

The size of the raster cell in a dataset is sometimes confused with the **minimum mapping unit**, i.e., the smallest element we can uniquely represent in our data. However, raster cell size and minimum mapping unit are *not quite* the same. Choosing an appropriate minimum mapping unit (or **resel**, an abbreviation for resolution element, as discussed in Chapter 1) for a study is a very important decision in the design phase of a project. Consider a dataset in which we wish to record vegetation types. In Figure 4.3a is a sketch map that might have come from a field mapping exercise. In the study area, a patch of evergreen trees is surrounded by grassland. One way to manually convert this data to a raster form is to overlay a regular grid on the sketch map (Figure 4.3b) and to assign a vegetation class to each cell, based on which class covers the majority of the cell.

Consider the problem shown in Figure 4.3c. A different grid with larger cells has replaced the first grid. Four raster cells overlay the small area where we have grassland surrounding the stand of evergreen trees. Based on the majority rule we chose above, the evergreen stand will not appear in the database. Each of four cells is covered by a small piece of the evergreen stand, but in each of these cells, the evergreen stand is only a minority component of the total cell area.

If the size of the raster cells had been different, the stand's existence would have been captured, as it was in Figure 4.3b. We might have even captured this stand if the cells in the raster had been oriented differently. This illustrates the need to understand the relationship between the minimum mapping unit and the raster cell size. A convenient rule of thumb, based on statistical sampling theory, is to use a raster cell half the length (or one-fourth the area) of the smallest feature you wish to record. A more conservative suggestion would be to use a raster cell one-third or one-fourth the length of the smallest desired feature.

To reinforce this important issue, the size of the resels in a dataset *must* be significantly larger than the size of the raster cells, or we run the risk of losing important spatial objects when we build the database.

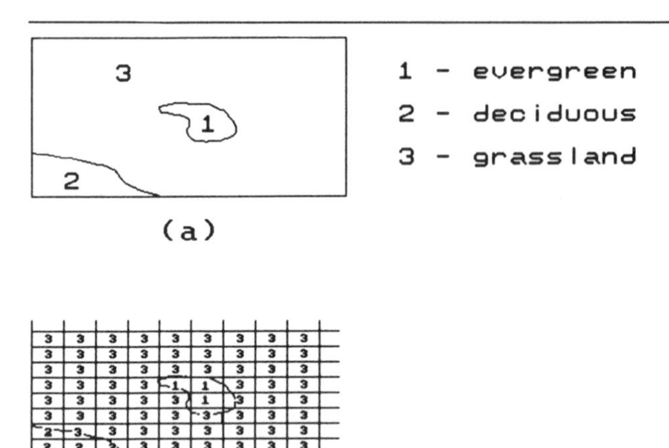

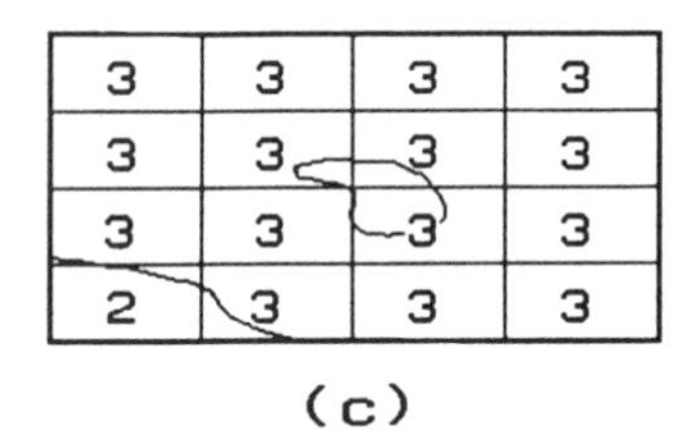

Figure 4.3 Minimum mapping unit. (a) vegetation map, (b) map converted to raster, (c) map converted to raster with poor choice of mapping unit.

Geometrical figures that completely cover a flat surface are called **tessellations.** The square raster structure we've just discussed is one such tessellation. Triangles and hexagons are two other tessellations of the plane (Figure 4.4). There has been a continuing interest in regular hexagonal cells as the basis of spatial data structures, in part because in a hexagonal tessellation of the plane, all neighboring cells are equidistant, unlike the situation in a raster of square cells (Burt, 1980). However, the use of a hexagonal or even a triangular data structure creates two problems that may be significant in some circumstances. First, the cells cannot be recursively subdivided into smaller cells of the same shape as the original cells, as is the case in a square system. Conversely, a hexagon made up of smaller hexagons will not be the same shape as those smaller hexagons. A third and less important point is that a numbering system for a hexagonal system is more complex than that of a square system, imposing at least a small additional overhead in system operations.

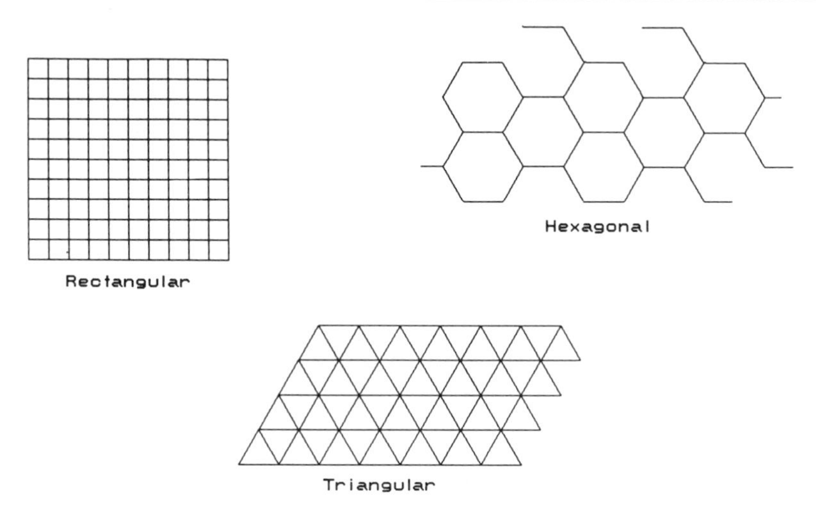

Figure 4.4 Tessellations of the plane.

Let's return briefly to the golf course example from the first chapter and create a raster database for this site. The simplest of the three datasets we considered was the map from the local planning agency, which describes the legal property boundaries, streets, and restrictions on construction and development due to easements for public utilities. In Figures 4.5a and 4.5b, respectively, we show a version of the original map, and a raster-converted representation of the map. The numbers in each cell indicate the permitted land use for each cell, using a majority rule as discussed above. By adding up the number of cells in each category, we can determine the fractional area coverage of each land use in the map:

Land-Use Category	Class	Total Cells	Percent of Total
Roads	1	213	39.0%
Easement Restrictions	2	129	23.6%
Unrestricted Development	3	204	37.4%

Raster datasets in practical use can be very large. As an example, satellite remote sensing data is frequently used to distinguish categories of land cover over large areas. The standard view of the earth from the U.S.'s Landsat series of satellites covers approximately 30,000 square kilometers,

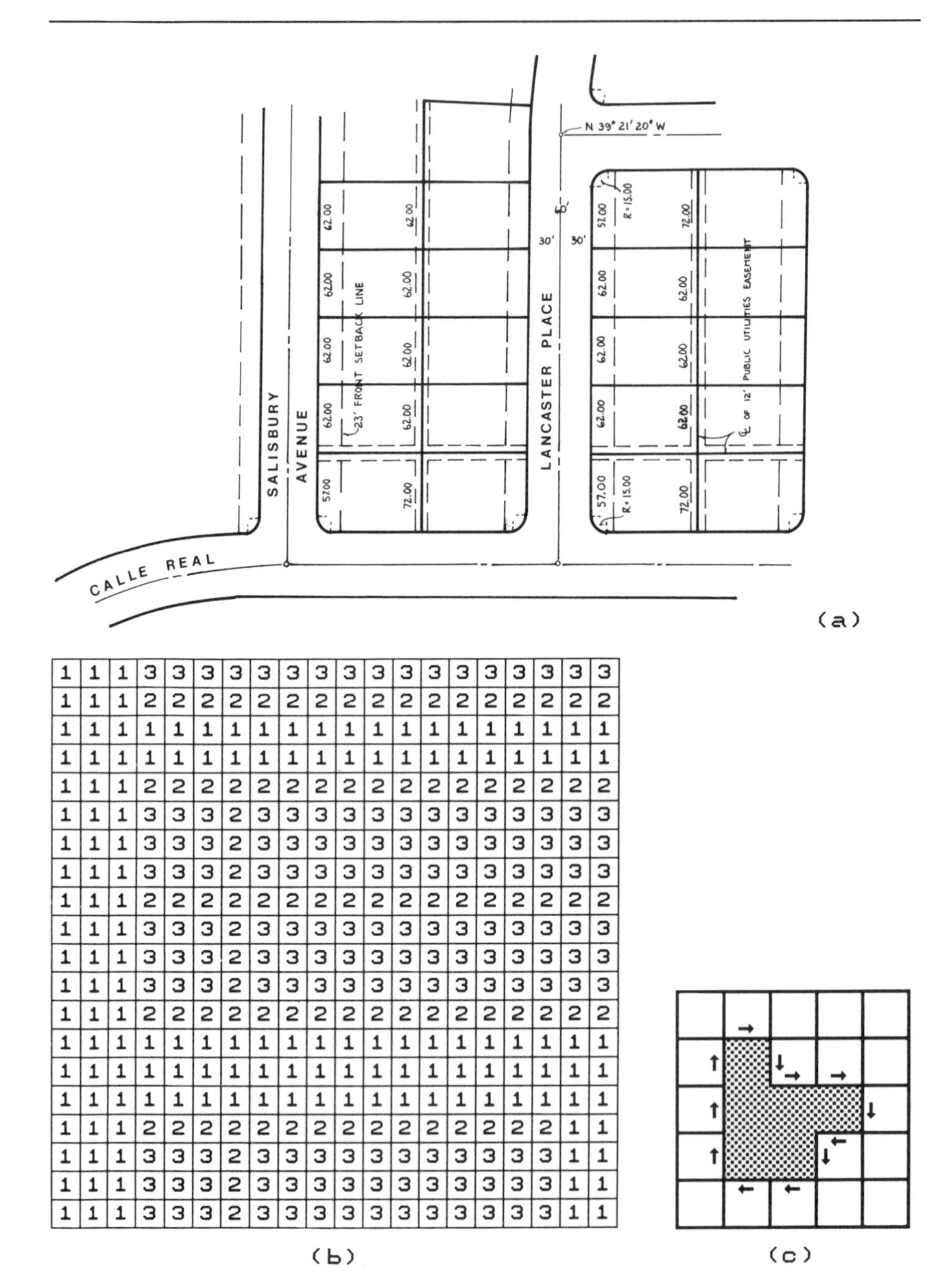

1	1	1	3	3	3	3	3	3	3	3	3	3	3	3	3	3	3	3	3
1	1	1	2	2	2	2	2	2	2	2	2	2	2	2	2	2	2	2	2
1	1	1	1	1	1	1	1	1	1	1	1	1	1	1	1	1	1	1	1
1	1	1	1	1	1	1	1	1	1	1	1	1	1	1	1	1	1	1	1
1	1	1	2	2	2	2	2	2	2	2	2	2	2	2	2	2	2	2	2
1	1	1	3	3	3	2	3	3	3	3	3	3	3	3	3	3	3	3	3
1	1	1	3	3	3	2	3	3	3	3	3	3	3	3	3	3	3	3	3
1	1	1	3	3	3	2	3	3	3	3	3	3	3	3	3	3	3	3	3
1	1	1	2	2	2	2	2	2	2	2	2	2	2	2	2	2	2	2	2
1	1	1	3	3	3	2	3	3	3	3	3	3	3	3	3	3	3	3	3
1	1	1	3	3	3	2	3	3	3	3	3	3	3	3	3	3	3	3	3
1	1	1	3	3	3	2	3	3	3	3	3	3	3	3	3	3	3	3	3
1	1	1	2	2	2	2	2	2	2	2	2	2	2	2	2	2	2	2	2
1	1	1	1	1	1	1	1	1	1	1	1	1	1	1	1	1	1	1	1
1	1	1	1	1	1	1	1	1	1	1	1	1	1	1	1	1	1	1	1
1	1	1	1	1	1	1	1	1	1	1	1	1	1	1	1	1	1	1	1
1	1	1	2	2	2	2	2	2	2	2	2	2	2	2	2	2	2	1	1
1	1	1	3	3	3	2	3	3	3	3	3	3	3	3	3	3	3	1	1
1	1	1	3	3	3	2	3	3	3	3	3	3	3	3	3	3	3	1	1
1	1	1	3	3	3	2	3	3	3	3	3	3	3	3	3	3	3	1	1

Figure 4.5 A categorical raster map. This map shows a raster structure for a land-use layer, based on the golf course design example. (a) subdivision map, (b) raster coded version of the subdivision map, and (c) chain code representation.

which at a nominal pixel size of 30 meters, corresponds to approximately 35 million raster cells or **pixels** (where pixel is a contraction of "picture element").

When dealing with such large datasets, there are several algorithms used to compress the data. Some of these algorithms are completely reversible; that is, we may recover exactly the original datasets. Others minimize the volume of the stored data by losing a (preferably) small and controlled amount of the original information. We will briefly mention two perfect-recovery compression mechanisms. The first, called **run-length encoding**, tries to exploit the fact that many datasets have large homogeneous regions. Consider the raster data in Figure 4.5b. The data attribute values at the beginning of the sixth row from the top are:

```
1  1  1  3  3  3  2  3  3  3  3  3  3  3  3  3
```

In a run-length encoded version, the original data is replaced by data pairs or **tuples**. The first number in the pair is a counter, indicating how many repetitions of the second number, the data value, occur starting at that point in the row. Thus, three cells in a row with data value 1 are compressed from three elements (1 1 1) to two (3 1). The data from the beginning of row 6 in our example would then become:

```
(3  1)  (3  3)  (1  2)  (9  3)
```

Thus, we have three 1s, followed by three 3s, followed by a single 2, and then nine 3s. In this case, the original data occupied 16 elements and the compressed data 8 elements, for a compression factor of 50%. Note that we have assumed that the data elements in the run-length encoded file - both attribute values and repeat counts - occupy the same amount of space. The effectiveness of this compression mechanism varies with the dataset. In the worst case, where there are no repeating sequences at all along the rows of the array, the algorithm will make the dataset twice as large. For binary data (in other words, data with only two possible classes), Burrough (1986) shows another kind of run-length coding, where in a single row, the position of a cell where a run begins is stored.

A second technique for compressing raster datasets uses what are called **chain codes**. In some instances we can consider a map as a set of spatially referenced objects placed on top of a background. The use of chain codes takes this point of view. The coordinates of a starting point on the border of an object (for example, a reservoir) are recorded, and then we store the sequence of cardinal directions of the cells that make up the boundary (Figure 4.3c). This may be an efficient means to store areas, particularly since each spatial object is kept as a separate entity in the database. However, some kinds of processing will require that the entire raster array be reconstituted which may be an unacceptable cost.

4.1.2 Hierarchical Raster Structures

There has been a great deal of interest recently in a family of enhancements to the standard square-celled raster structure. We will introduce these modifications through an example. Consider a set of digital elevation data values, where the fundamental data are stored on a 50-meter square grid (that is, each cell represents a square that is 50 meters on a side). Rather than storing this information as a single layer in our GIS, we shall store it in several interrelated layers. One layer corresponds to the original 50-meter interval raster data. A second layer consists of data resampled to a 100-meter interval. Each cell in the 100-meter layer is the algebraic average of four cells in the 50-meter layer (Figure 4.6a). A third layer is created by averaging four 100-meter cells to create 200-meter cells. And we could continue this spatial averaging process, decreasing the spatial resolution at each "higher" layer, until at the highest layer we might have a single pixel, whose elevation value is the numerical average of all the data in the original 50-meter layer. As an aside, this only works perfectly - that is, yields a single pixel in the highest layer - if the original 50-meter data was a square array of 2^n pixels on a side.

In general, this is called a **pyramidal** data structure, since we can imagine each of the derived layers stacked on top of previous layers, in the shape of a pyramid. If each higher level layer has pixels that are exactly twice as wide as the previous (and thus, four times the area), as in our example, this is called a **quadtree** data structure (Samet, 1984). The name *quadtree* comes from the four-fold reduction in number of pixels in each layer, and the fact that the structure is easily pictured (as well as represented in the digital database) as a tree (Figure 4.6b). Leaves at the bottom of the tree, in this case, represent the elevation values in the original 50-meter data layer, and

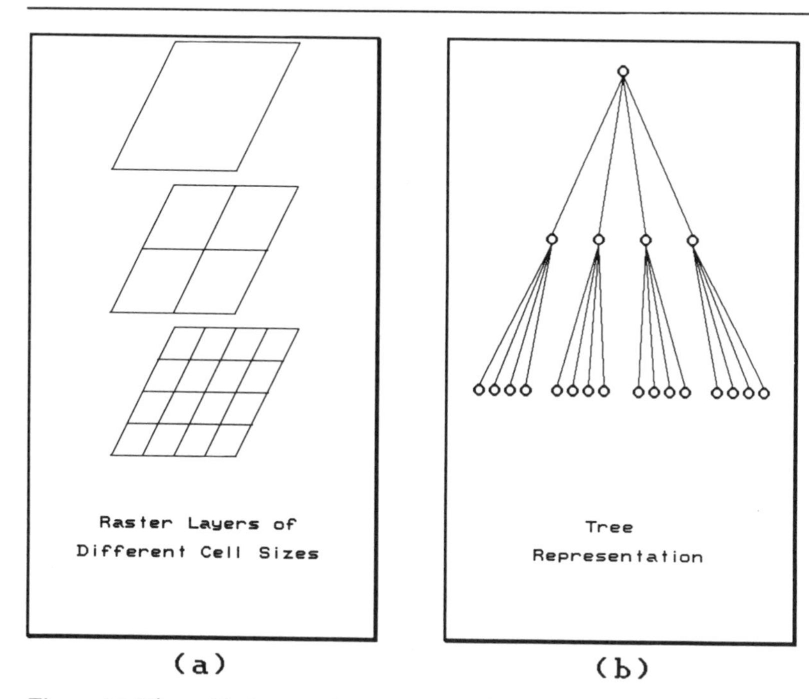

Figure 4.6 Hierarchical raster data structures. (a) pyramid, (b) pyramid drawn as a tree.

branches in the tree represent the averaging process as connections between pixels or cells in the different layers.

Tobler and Chen (1986) discuss a modified quadtree system that may be useful for coding data for the entire Earth's surface. The single node at the top level of the tree represents the entire planet. At the 15th level, the resolution of the cells is comparable to that of meteorological satellites. At the 26th level, the spatial resolution is comparable to most aerial photography, and at level 30 there is centimeter-scale resolution over the globe, which Tobler and Chen indicate is adequate for geodetic control points in many applications.

Data stored in a simple quadtree will occupy somewhat more storage space than the original raster. This is true because the original raster is a component of the pyramid, and forms the lowest layer in the quadtree. Consider a raster with 32 pixels on a side. This raster requires (32 times 32) or 1024 cells be stored. If we include the higher level layers, we require:

Layer	Width in Cells	Total Cells
1	32	1024
2	16	256
3	8	64
4	4	16
5	2	4
6	1	1

which totals 1365 cells. For this small example, we increase the storage overhead 33%, since there are 33% more cells in the complete tree than in the original raster.

However, this storage cost can be offset by some distinct advantages. There may be some processing steps that do not require the full resolution of the original raster dataset. If the data have been stored in a quadtree, a given step in the analysis may be based on whatever layer in the quadtree holds information of the appropriate resolution or scale, potentially saving both computer system input/output and processing time. A situation in which this advantage may be significant is where different datasets in the geographic information system have different mean resolutions. This is one of many examples we will discuss of the classic database management trade-off of storage costs versus processing costs, which we will see again in our discussions.

Additionally, the hierarchical nature of the quadtree may permit us to minimize some kinds of search through the database (Smith et al., 1987). For example, if we are looking for a high elevation in the database (but we don't necessarily need the *highest* elevation), we could instruct our system to work *down* the quadtree: first find the quadrant in the next-to-uppermost layer with the highest average elevation, and ignore the rest of the data. In Figure 4.7, the lower-right quadrant of the two-by-two layer of the quadtree is the highest, with an elevation of 121 meters. We therefore consider only this fraction of the data at the next level down, removing 75% of the data from further consideration.

Continuing down towards the highest-precision data layers in Figure 4.7, we find the highest cell in the lower-right corner of the four-by-four layer at an elevation of 127 meters. Repeating the procedure in the next level of the quadtree, we locate the cell marked as 134 meters. In total, we have examined 12 cells, out of total database of 64 elemental cells at the bottom of the tree, or 84 cells in the complete tree. In this way, we can search only a small fraction of the entire database, and still find an acceptably high place for many purposes. However, with this search strategy there is no guarantee that we will find the highest location in the database.

If we stored more than just a single data value at each node in the quadtree, we can even satisfy a specific search for the highest location in the database. Consider a version in which we store three values at each quadtree node: the mean elevation of the appropriate area, as well as the minimum and maximum known elevation values in this area. In this case, we can use a search strategy parallel to the one discussed above to be able to find the highest (or lowest) cell in the entire region in a very small amount of time. This is another example of the compromise between storage and processing. Here we have additional storage costs (since we store three values in each cell) in order to decrease the time required to search the database.

Another way in which a quadtree data structure can be used to minimize search effort involves categorical data. Imagine a quadtree for storing a number of land-use classes. At the base of the tree, each cell takes the attribute for the dominant class in the cell. However, each composite cell at the each higher level of the tree does not have to store just the dominant class for the cell or just the fact that the cell is not homogeneous; the higher-level cells are not limited to storing a single value. Instead, each higher-level cell could store a list of all the land-use classes stored in the cells beneath it.

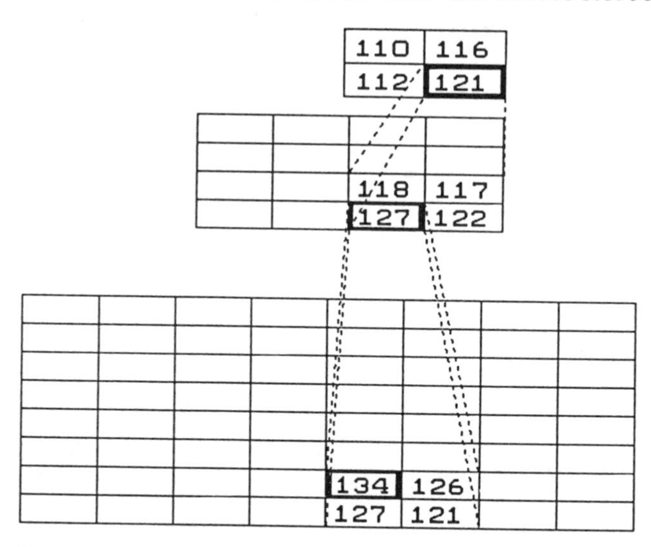

Figure 4.7 Efficient search via the quadtree structure.

In this way, when we search through the quadtree looking for areas with certain kinds of land use, going from less geographic detail to more, we can discard limbs of the tree with no error.

A modification of the quadtree can be used to minimize system storage under some circumstances. In a **maximum block representation**, we systematically eliminate all the redundant information in the tree. Consider the land-use example in Figure 4.8, where each cell in the original raster (Figure 4.8a) is coded in a binary fashion for land use. In this example, 0 represents undeveloped land surrounding a city, and 1 represents urban land use. In the associated quadtree, open circles represent the undeveloped land, and solid boxes represent the urban land use. For each lower-resolution level, we average the pixel values in the higher-resolution layer, to code for the fractional coverage of urban area in a cell.

The numbering scheme to specify cells in a quadtree proceeds from the upper left in the data array; cell numbers are recorded in the upper right corner of the individual cells in the accompanying figures. Starting with the 2-by-2 area in the upper left, cells 1 through 4 are the upper left, upper right, lower left, and lower right, in sequence. The 2-by-2 area in the upper right is then numbered in the same pattern, then the area in the lower left, and finally we number the cells in the lower right. In a larger data array, we would then number the cells in the upper right corner 4-by-4 array, and so on.

However, notice that there is redundancy in the companion quadtree to this raster array (Figure 4.8b). Cells 5, 6, 7, and 8 are all class 0, thus their average is zero; therefore specifying the exact attribute value of zero in the higher level layer tells us everything about the coordinating pixels in the lower-

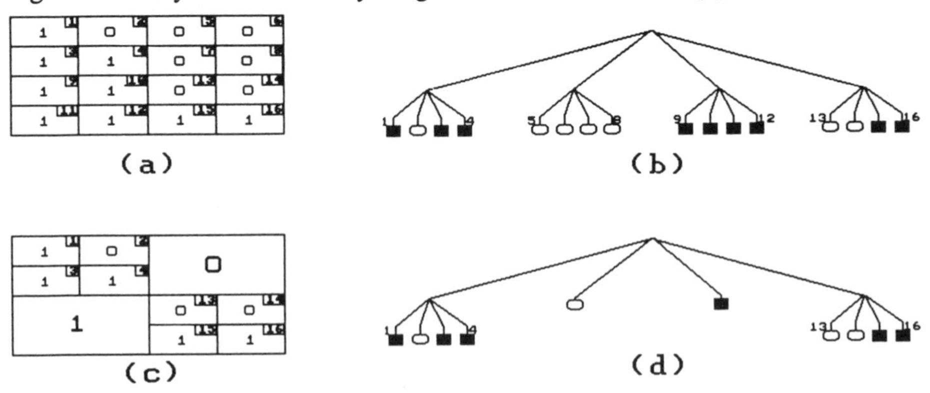

Figure 4.8 Quadtree maximum block representation. (a) land use, (b) land use as a tree, (c) maximum block description, (d) equivalent tree.

level layer. In this case, we could leave out the nodes in the tree at the lower (or increased resolution) level, thus decreasing the storage costs without losing any information. This is the maximum block representation (Figures 4.8c and 4.8d) of the quadtree. In this case, the maximum block quadtree requires eight fewer nodes to describe the spatial arrangement of land use. This revised tree requires a bit more processing to construct and use than a simple quadtree, since one must test to determine whether a cell at a particular resolution exists as a unique entity, or is represented as a part of a composite larger-area node in a different layer.

One problem with the quadtree data structure is that it is not invariant under translation, rotation, or scaling. This is the same problem that is found with any arbitrary subdivision of datasets. Arbitrary boundaries are placed on the underlying continuous data to develop the quadtree, and the locations of the boundaries can strongly affect the resulting derived data. Consider a simple example of translation along one of the major axes of the raster array. Figure 4.9a shows a simple object composed of six shaded raster cells, along with the maximum block quadtree representation. The shaded cells in the raster array are indicated in the tree by filled squares; the white background is

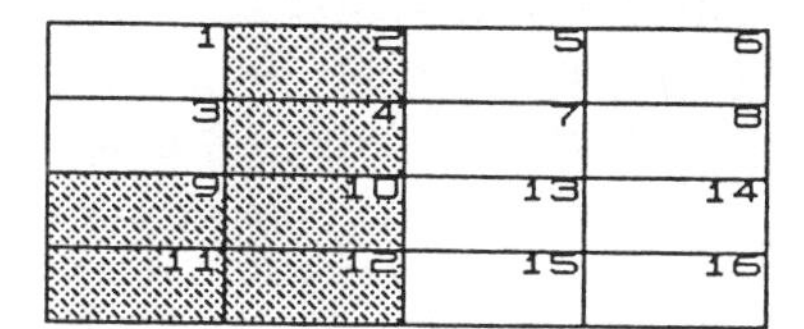

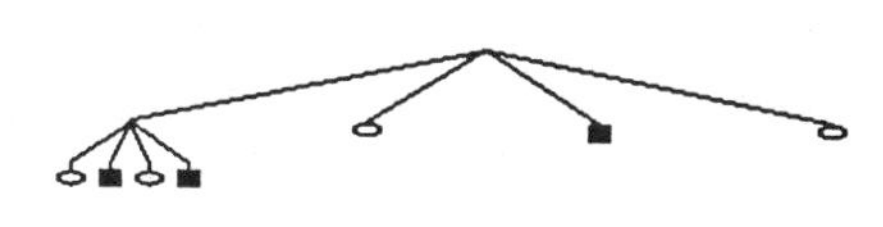

Original Raster and Quad-Tree

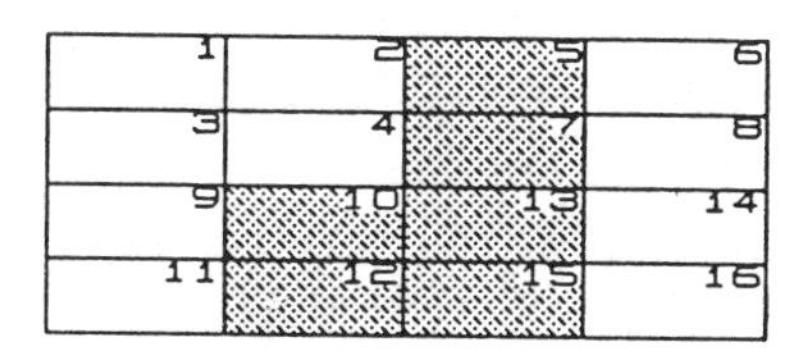

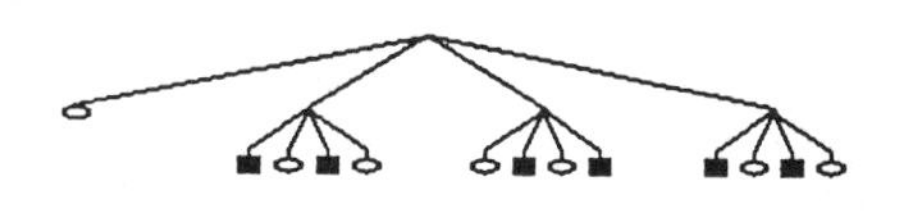

Translated Raster and Quad-Tree

Figure 4.9 Quadtree translation. The shaded blocks in the array correspond to the filled cells in the tree, and the open cells in the array correspond to the open circles in the tree. While the shape of the object in the two arrays are the same, the shape of the comparable quadtrees are not.

indicated in the tree by open circles. What happens when we translate the object to the east by one cell width? This could represent either another object of the same kind in a different location, or another quadtree with the same cell size but a different starting location. The object in Figure 4.9b looks the same to the human interpreter, but evolves into a quadtree with a very different shape. One solution to this problem is offered by Scott and Iyengar (1986), who discuss a related data structure that is translation invariant - that is, if the entire figure is moved, the characteristics of the data structure will not change. Their system is based upon (1) recognizing homogeneous square regions in a raster, and (2) recording both the size of the block and the coordinates of the block's upper left-hand corner.

4.2 Vector Data Structures

A mathematician might define a **vector** as a quantity with a starting coordinate, and an associated displacement and direction (or bearing). In a description of spatial data based on vectors, we make the assumption that an element may be located at any location, without the positional constraints of a raster array. To briefly review the previous section, raster data structures are based on (usually regular) decompositions (i.e., "breakdowns") of the plane. In raster-based systems for representing spatial data, our ability to specify a location in space is limited by the size of the raster elements, since we are unable to know anything about different locations within a raster cell. In other words, there is a limit to geographic specificity.

Vector data structures are based on elemental points whose locations are known to arbitrary precision, in contrast to the raster or cellular data structures we have described. As a simple example, to store a circle in one of the raster data structures above, we might find and encode all the raster cells (of a pre-defined shape and size) whose locations correspond to the boundary of the circle. This might be called a low-level description of the circle. A high-level description, on the other hand, might efficiently store the circle by recording a point location for the center of the circle, and specifying the radius. In this example, note that the high level description, based on a vector representation, is more efficient in terms of the amount of data required, as well as more precise, if we have a means to indicate the geometric object **circle**. Many computer graphics and computer-aided-design (CAD) systems use vector-like models as their internal data organization, using primitives such as points, lines, and circles. These elements may also be found in the computer graphics language standards, such as GKS (Enderle et al., 1984), as well as many personal computer languages such as BASIC. These advantages may disappear, however, if we have to store the circle as a connected sequence of straight-line segments.

For spatial data in most geographic information systems, the coordinate data is encoded, and after input processing, is stored as some combination of points, lines, and areas or polygons (Males, 1977; Peucker and Chrisman, 1975; and Peuquet, 1977). Several forms of vector data structures are in common use, both as representative database types within geographic information systems, as well as standards for data transfer between systems:

- whole polygon structure
- Dual Independent Map Encoding (DIME) file structure
- arc-node structure
- relational structure
- digital line graphs

We will discuss each of these in turn.

4.2.1 Whole Polygon Structure

In a **whole polygon structure**, each layer in the database is divided into a set of polygons (Figure 4.10). Each polygon is encoded in the database as a sequence of locations that define the boundaries of each closed area in a specified coordinate system (sometimes called a boundary loop). Each polygon is then stored as an independent feature. There is no explicit means in this system of referencing areas that are adjacent. This is, to some extent, comparable to the chain-coded raster discussed in the previous section: in both whole polygon structure and a chain coded raster, the emphasis is on the individual polygonal areas, where each discrete area is stored separately. Attributes of the polygons, such as land cover or ownership, may be stored with the coordinate list. By maintaining each polygon as a separate entity in this way, the **topological organization** of the polygons is not maintained. By topology, we refer to the relationships between different spatial objects: which polygons share a boundary, which points fall along the edge of a particular polygon, and so forth. In a whole polygon structure, line segments that define

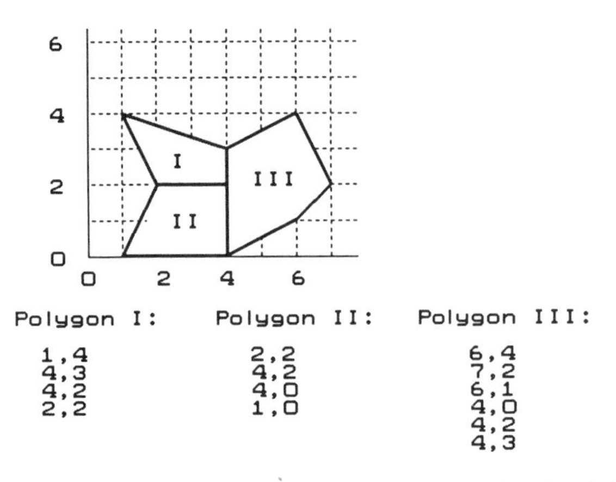

Figure 4.10 Whole polygon structure. The nodes that define each polygon are stored separately.

the edges of polygons are recorded twice - once for the polygon on each side of a line. Similarly, points that are shared by several polygons, such as location (4,2) in the example, will also be represented several times in the database. With this organization, editing and updating the database without corrupting the data structure can be difficult.

4.2.2 DIME Structure

The Dual Independent Map Encoding or **DIME** file structure was developed for use by the U.S. Bureau of the Census (Cooke and Maxfield, 1967). It was designed to incorporate topological information about urban areas for use in demographic analyses (Cooke, 1987). While DIME files themselves do not generally correspond to the internal database organization of a GIS, they are in common use as an archive data format as well as a defined format for data exchange between different systems. The basic element of the DIME file structure is a line segment defined by two end points or nodes. Figure 4.11 illustrates the structure of the DIME file, slightly simplified.

The line segments and nodes are shared by adjacent polygonal units. In this structure the line segments are assumed to be straight. When curved lines are needed, they are represented as sequences of straight line segments. Each line segment is stored with three essential components: a segment name (such as the name of the street) that identifies the segment, node identifiers for the

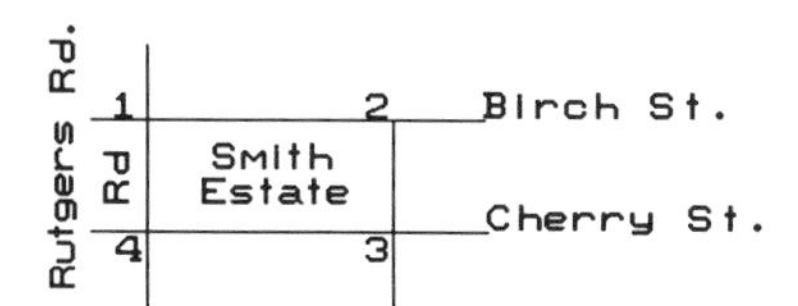

Header Items: Groups of Segments

Header Number	Zip Code	Area Code	Tract
1000	93106	805	14
1001	93117	805	14

Segment Codes: Each Line Segment

Segment Name	Nodes From	Nodes To	Polygon Left	Polygon Right	Addresses Left Low	Addresses Left High	Addresses Right Low	Addresses Right High	Header
Birch St	1	2	-	Smith Est	101	175	102	178	1000
Cherry St	3	4	-	Smith Est	103	177	104	180	1000
Rutgers Rd	4	1	-	Smith Est	8602	8686	8603	8685	1000

Node Locations:

Node	Easting	Northing
1	127,251	1,340,600
2	127,352	1,040,601
3	127,350	1,040,584
4	127,256	1,040,502

Figure 4.11 DIME file structure.

"from" and "to" endpoints of the segment, and identifiers for the polygons on left and right sides of the segment.

A number of additional attributes may be coded in the DIME file structure, in order to store additional information about the various spatial objects. When the segment is a part of street, the address ranges for both sides of the street may be stored. A field is available for segments that are not streets, to indicate such features as a proposed street or the shoreline of a lake. Additional attribute fields, labeled by header numbers, are available for groups of segments, such as a telephone exchange, mailing address code (such as ZIP codes in the United States or similar mailing codes in the United Kingdom and Canada), or voting districts. There is a separate set of fields kept for logically grouping and labeling segments. In the example in the figure, the correspondence between headers and the mailing and telephone exchange codes are represented.

A major disadvantage of DIME structures lies in the difficulty of manipulating complex lines, as in functions that require search along streets.

Since streets are broken into discrete street segments by the cross streets, it is a significant computational effort to follow the segments in sequence when required. An advantage of the system for some applications is its ability to match addresses of spatial objects in multiple files, since the addresses are explicitly stored in the DIME file.

4.2.3 Arc-Node Structure

In an **arc-node** data structure, objects in the database are structured hierarchically. In this system, points are the elemental basic components (Figure 4.12). Arcs are the individual line segments that are defined by a series of x-y coordinate pairs. Nodes are at the ends of arcs and form the points of intersection between arcs. There may be a distinction made between nodes at the ends of lines, and points that are not associated with lines (as we did in our discussions earlier). Polygons are areas that are completely bounded by a set of arcs. Thus, nodes are shared by both arcs and contiguous polygons (Peucker and Chrisman, 1975). Several commercial geographic information systems use forms of this arc-node data structure.

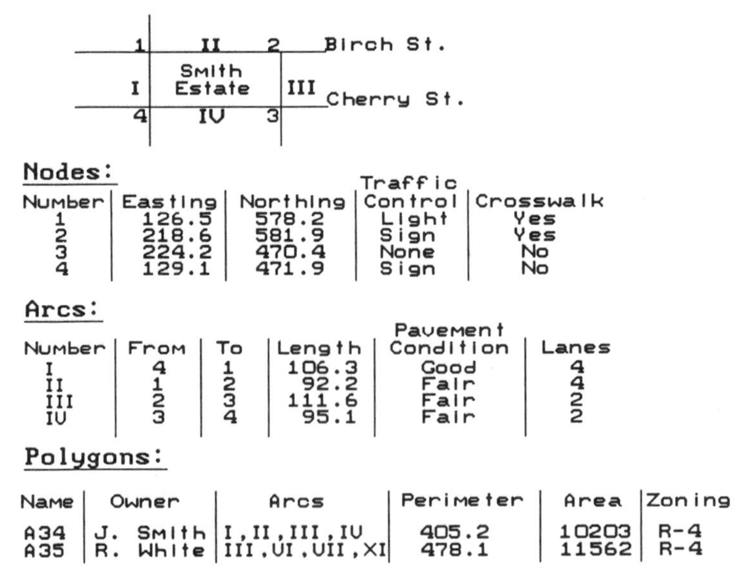

Nodes:

Number	Easting	Northing	Traffic Control	Crosswalk
1	126.5	578.2	Light	Yes
2	218.6	581.9	Sign	Yes
3	224.2	470.4	None	No
4	129.1	471.9	Sign	No

Arcs:

Number	From	To	Length	Pavement Condition	Lanes
I	4	1	106.3	Good	4
II	1	2	92.2	Fair	4
III	2	3	111.6	Fair	2
IV	3	4	95.1	Fair	2

Polygons:

Name	Owner	Arcs	Perimeter	Area	Zoning
A34	J. Smith	I, II, III, IV	405.2	10203	R-4
A35	R. White	III, VI, VII, XI	478.1	11562	R-4

Figure 4.12 Arc-node data structure.

Arc-node structures permit us to encode the geometry of the data with no redundancy. Unlike the whole polygon structure, points are stored only once. They are reused as often as necessary, as in Figure 4.12 where point 1 is a component of polygon A as well as lines I and II. In an arc-node database, we can easily include attributes of various kinds. In the street and traffic control example in Figure 4.12, attributes are explicitly linked to geometry. For example, traffic control device descriptions are stored with the relevant nodes, and roadway length and pavement condition are stored with the appropriate arcs.

Notice that in this arc-node transportation example we have stored the lengths of the arcs. This was a conscious decision that adds some redundancy to our database, since we could in theory calculate these lengths from the coordinates of the nodes at the arc ends. If our application makes intensive use of arc lengths, it may be more efficient to determine the lengths once when the database is constructed, rather than recalculating this value many times as the data is used. This is another example of the classic problem in database management systems of achieving a balance between storage and retrieval costs in comparison to processing costs. In a modern GIS, the user should be able to affect this balance, based on specific applications requirements.

4.2.4 Relational Structure

Another form of arc-node vector data organization is sometimes called **relational data structure**. In the last arc-node example, data attribute values were stored together with topological information. In a relational data structure, attribute information is kept separately. This has become a popular design strategy for several commercial geographic information systems. We'll recast the same data from in the arc-node example of Figure 4.12 into this alternative organization.

The topological data in a relational structure is organized in the same fashion as in an arc-node organization. The principal difference is in the attribute values. These are stored in relational tables (as in a relational database management system, which is discussed in more detail in section 7.3.1) as in Figure 4.13. These tables are straightforward: a row in a table represents a single data record, and the columns represent the different fields or attributes. One of the relational tables keeps track of the attributes of points or nodes in the spatial database. In our example, this table records

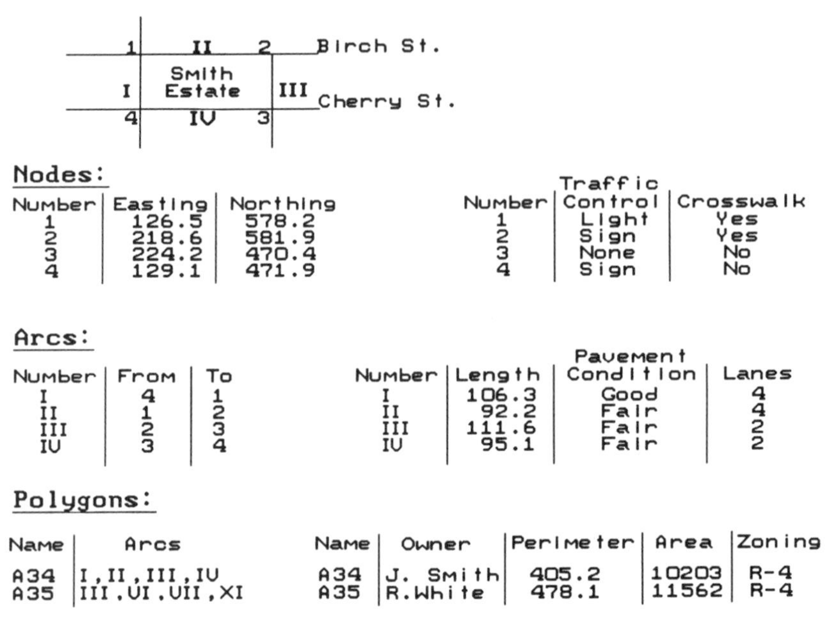

Nodes:

Number	Easting	Northing
1	126.5	578.2
2	218.6	581.9
3	224.2	470.4
4	129.1	471.9

Number	Traffic Control	Crosswalk
1	Light	Yes
2	Sign	Yes
3	None	No
4	Sign	No

Arcs:

Number	From	To
I	4	1
II	1	2
III	2	3
IU	3	4

Number	Length	Pavement Condition	Lanes
I	106.3	Good	4
II	92.2	Fair	4
III	111.6	Fair	2
IU	95.1	Fair	2

Polygons:

Name	Arcs
A34	I,II,III,IU
A35	III,UI,UII,XI

Name	Owner	Perimeter	Area	Zoning
A34	J. Smith	405.2	10203	R-4
A35	R.White	478.1	11562	R-4

Figure 4.13 Relational data structure.

pointers to the nodes, the kind of traffic control that exists at this node, and whether there are marked crosswalks at the street intersection. Similar tables for arcs might hold information about road lengths, pavement type, and the dates the street had been repaved. The tables for polygons might record area, zoning, assessed value, and ownership.

Notice that this manner of storing data is directly comparable to that of the arc-node format in the previous section. The main difference is that, in the relational structure, the attribute data are maintained separately from the topological information. Thus, there are more separate files and pointers to maintain. More detail on the relational database model is found in section 7.3.1. General-purpose database management software is available for the relational model. Geographic information systems that take advantage of this software may cost less to develop and may provide added flexibility, by relying on a commercial database management system.

4.2.5 Digital Line Graph Structure

The U.S. Geological Survey has made tremendous investments in digital cartographic data, and in the process, has developed a number of standards for such data. One of the best known is a standard vector file format known as a **digital line graph** (DLG). The agency is producing these vector files based on the source materials used to compile the USGS 7.5-Minute and 15-Minute Topographic Maps Series. A separate set of DLG data files is based on 1:2,000,000 scale map products. The data contents of the DLG files are subdivided into different thematic layers (Allder and Elasal, 1984). One layer consists of boundary information, including both political and administrative boundaries in the region. A second layer is for hydrographic features. A third layer is for the transportation network in the area. Finally, the fourth layer is based on the Public Land Survey System, which has as its focus a survey system administered by the U.S. Bureau of Land Management.

The essential data elements of the DLG level 3 structure are similar to the other vector data structures discussed in sections 4.2.2, 4.2.3, and 4.2.4. Nodes represent either end points of lines or line intersections, while additional points are used where required to indicate significant features along lines. Lines have starting and ending nodes, and as such, permit us to specify a direction along the line as well as the areas on the left and right sides of the line. A special **degenerate line** is defined as a line of zero length, and is used to define features that are indicated on the map as a point. Degenerate lines are recognizable because they have the same starting and ending node. Areas in the DLG format are completely bounded by line segments. Each area may have an associated point that represents the characteristics of the area; the point location is arbitrary, and may not even be within the area.

The point, line, and area elements provide information about topology and location. In addition, there is an extensive system for coding attribute information for the elements. The attribute codes are based on those features represented on USGS topographic maps. Attribute codes are structured in a specified way, with both major and minor code components. Allder et. al. (1984) provide the details of the attribute coding. The major code is three digits long. The first two digits denote the general category of the element, while the third digit provides additional detail. A few of the major code categories are:

Major Code	Category
020	Hypsography
050	Hydrography
070	Surface Cover

Minor codes are four digits long. The first digit is generally zero. The remaining three digits provide further detail:

Minor Code	Description
001 - 099	nodes
100 - 199	areas
200 - 299	lines
300 - 399	degenerate line
400 - 499	general-purpose codes
600 - 699	descriptive codes

The general-purpose codes are used for those features that may be digitized as a node, an area, or a line, depending on the feature's size and position. The descriptive codes are used to provide additional information when required.

Several attribute codes have special meaning. Code 000 0000 is reserved for the area outside a given map sheet; remember that the DLG data series are based on topographic maps. Other codes are reserved for such information as features that have been revised photographically, and those features that cannot be identified from the available source materials. To illustrate some of the detail that may be stored in the DLG format, we present a few of the codes from the hydrography DLG data layer. Note that the 050 major code specifies hydrography.

Nodes
050 0001 Upper end of stream
050 0004 Stream entering water body
050 0005 Stream leaving water body

Areas
050 0101 Reservoir
050 0103 Glacier
050 0106 Fish hatchery

Lines
050 0200 Shoreline
050 0201 Man-made shoreline

Degenerate Lines
050 0300 Spring
050 0302 Flowing well

General Purpose Attributes
050 0400 Rapids
050 0401 Falls
050 0406 Dam or weir

General Descriptive Attributes
050 0601 Underground
050 0603 Elevated
050 0604 Tunnel

A DLG level 3 data file contains a number of header records, followed by the data records. The header records provide information about the date of creation of the file, map projection and coordinate system, and the number of points, lines, and areas stored in the file. Data records for nodes include a description (including the node location), the major and minor attribute codes, and a text string. Data records for areas include a description (including the coordinate location of the representative point), the attribute codes, and an associated text string. Data records for lines include a description (including identification of the starting and end nodes and the areas on the left and right), an ordered sequence of x-y coordinates along the line, the attribute codes, and a text string.

There is an optional final record with information about estimated accuracy of the data. Allder et al. (1984) present greater details of the DLG format, as well as example data files. We point out that there is a new development that is expected to supplant the DLG and DIME file structures for use in the 1990 U.S. Census, called TIGER: the *Topologically Integrated Geographic Encoding and Referencing* system. A detailed discussion of TIGER is outside the scope of this section.

4.3 Comparisons between Data Structures

A data storage structure may or may not incorporate **topological** information, describing not only an object's position, but also its spatial relationships with respect to neighboring objects. Topological information is important for many kinds of analyses, including automated error detection, windowing for analysis and graphic presentation, network applications, determining whether a point falls in a specific polygon, proximity operations, polygon overlay, and other intersection procedures (Peucker and Chrisman, 1975). However, if your application does not require this kind of detailed information about the relationships between spatial objects, the additional overhead of explicitly treating topology may significantly complicate the tasks of database creation and update. For example, unstructured vector lists may be perfectly adequate for some kinds of routine data display.

Some kinds of topological information are implicit in spatial data. In a simple raster-structured data file, for example, there is a specified spatial organization for the data, with no gaps in the fabric of the database. The regularity in the array provides an implicit addressing system. This permits rapid random access to specified locations in the database. Thus, we know immediately those cells that are adjacent to any target location, and we can easily find and examine those regions that bound a specified group of cells. Topological information in vector structures is often coded explicitly in the database. Line segments within DIME files, for example, have identifiers and codes for the polygon on either side. When needed topological relationships are not explicitly coded in vector data structures, it can be relatively expensive and time-consuming to get the system to come up with them. For example, in some operational systems, data entry from maps starts with a digitizing process that does not require that the operator explicitly relate various line segments and polygons that use these segments as parts of their boundaries. Instead, after the operator creates what is sometimes called *cartographic spaghetti*, a batch process links points, lines, and polygons together into a topologically structured database. This is another illustration of the trade-off between the efforts expended when developing a database and additional storage costs versus costs and speed during analysis and retrieval.

The traditional advantages and disadvantages of raster versus vector spatial data structures have been documented by Kennedy and Meyers (1977) and Durfee (1974), among many others. Basic issues include data volume (or storage efficiency), retrieval efficiency, robustness to perturbation, data manipulation (or processing) efficiency, data accuracy, and data display. Some of these differences, however, are less important in modern GIS implementations. Comparisons of data volume between raster and vector systems is entirely dependent on the database contents, as well as considerations of accuracy and precision. Some may even argue that it is inappropriate to compare the two in terms of data volume, since the characteristics of the two are so different. For example, an elevation dataset is generally stored as a complete cellular array in a raster, and as line segments storing the locations of lines of constant elevation (i.e., the contour lines) in a vector model. These are fundamentally different views of the underlying spatial information: the raster is quasi-continuous, the vector is clearly discrete; the raster representation may be considered more *dense* than the vector because more unique values are stored.

Comparisons of processing efficiency are difficult in modern systems as well. Traditionally, overlay operations are thought to be more efficient in a raster system (see section 8.1.1). In current systems, there may be an efficient means to determine the approximate locations of polygons by maintaining a separate index database. Using such an index to structure a search through the spatial data, a comparison of raster and vector data structures based on processing speed may be more sensitive to the spatial data itself than to the choice between the two data structures.

Durfee (1974) and others have suggested that modern geographic information systems should be able to accept, store, manage, retrieve, and display both raster and vector data types, as well as convert from one data structure to the other (Kennedy and Meyers, 1977 and Cicone, 1977). Unfortunately some existing GIS implementations do not have this capability.

If forms of both raster and vector structures are found in a GIS, as well as structure conversion routines and appropriate analysis tools for each data type, then the data could be stored in their "natural" form to both optimize geographic specificity and minimize conversion costs and attendant bias. This also permits analytic procedures to operate on a data structure where efficiency or accuracy are highest. While this strategy is more complex than one in which all data are stored and manipulated in a single data structure, efficient software and hardware for vector/raster conversions reduce the size of the problem.

The quadtree, as a more sophisticated means of working with raster data, may provide significant performance improvements over simple methods of raster storage. In exchange for increased complexity of the software to store and retrieve the data elements, as well as increased costs at the start to create the tree structure, there can be tremendous improvements in search speed, as we have discussed. A number of research groups, as well as commercial firms, are developing geographic information systems based on the quadtree data model. In particular, merging hierarchical raster data structures with vector data structures may take advantage of the benefits of each. We briefly examine one of these research systems in Chapter 12.

The performance of a pure arc-node database varies with the application. When you have retrieved a geometric entity (point, line, or polygon), you have also retrieved all the relevant attributes. If the attributes of each geometric entity are required, no additional database operations are required. On the other hand, there is no way to work with the geometrical entities without incurring the overhead of moving through the attribute data as well, since they are in a common database. This may exact significant performance penalties during the retrieval process.

The relational data structure has the potential for efficient search, at the expense of data file management complexity. As we have seen, such a system design permits search through either the geometrical entities or the attribute data, without the other getting in the way, since these two kinds of information are stored separately. Thus, one expects better data retrieval performance for simple kinds of search, which should result in more efficient operations. In any event, such a system can minimize the computer's input/output operations, at the expense of more complex file management, which may be particularly important when working on multi-user systems.

Chapter 5

Data Acquisition

The point is often made that the value of a geographic information system is due in large part to the quality of the data contained within the system. In this chapter we address the problem of acquiring the wide variety of essential datasets.

5.1 Introduction

The first steps in developing the database for a geographic information system are to acquire the data and to place them into the system. GISs must be able to accept a wide range of kinds and formats of data for input. There may be times when a given user may generate all their own datasets; this is, however, relatively rare. Even so, getting data is one of the greatest operational problems and costs in the field. Kennedy and Guinn (1975) described the importance of the data in automated spatial information systems in this way:

While models which use the data are important to support the decision-making activities of those who use the system, a large portion of the investment will be in obtaining, converting, and storing new data.

Data to be input to a GIS are typically acquired in a diverse variety of forms. Some data come in graphic and tabular forms. These would include maps and photography, records from site visits by specialists from many fields, related non-spatial information from both printed and digital files (including descriptive information about the spatial data, such as date of compilation, and observational criteria). Other data come in digital form. These would include digital spatial data such as computer records of demographic or land ownership data, magnetic tapes containing information about topography, and remotely sensed imagery.

Often these data will require manual or automated preprocessing prior to data encoding. For example, tabular records may need to be entered into the computer system. Aerial photography might require photointerpretation to extract the important spatial objects and their relative locations, a digitizing process to convert the data to digital form, and numerical rectification algorithms to convert the locations of significant features to a standard georeferencing system. Computer programmers may need to help move digital datasets from one computer to another. Airborne scanner data might require thematic classification and rectification before the data are suitable for entry into a GIS. We discuss many of these processes in Chapter 6. An important and sometimes overlooked issue during data acquisition is to obtain information about the accuracy, precision, currency, and spatial characteristics (such as the georeferencing system and scale) of the data themselves (Kennedy and Meyers, 1977; and Salmen et al., 1977). In some cases, there may be mapping standards of various kinds, which will vary by country, organization, and agency.

We can think of a number of broad classes of data to guide our discussions of data acquisition. Many interesting spatial datasets are effectively sets of points. These include the locations of water, gas, and oil wells, a representative location for groups of buildings, and the addresses of members of a target demographic group. Other datasets may be considered forms of networks. The complete set of roads in an area, ranging from unpaved fire trails to multi-lane superhighways, is one kind of transportation network. Other networks include waterways, potable-water delivery and waste-water collection systems, railroad systems, and gas, electric, and communications systems at local and regional levels.

Other kinds of spatial data might be described as continuous fields, where we can theoretically calculate or measure a value at any location. Examples of this class of data are descriptions of elevation, plant biomass, or population density (and many kinds of demographic variables). Finally, an important class of spatial data involves dividing a portion of the Earth's surface into relatively homogeneous discrete regions. A political map is perhaps the most common example of this type, since it subdivides a portion of the Earth into countries, states, provinces, counties, and so forth. A similar way to subdivide the Earth would be to develop classes of land cover or land use, indicating the boundaries of the different classes and the characteristics within the boundaries.

Each of these different kinds of spatial data may be stored and presented to users in various ways. Traditional cartographic products provide an easy-to-understand means of storing and communicating these various kinds of spatial data. In many cases, photography may be used, perhaps in combination with cartographic overlays, to portray phenomena and relationships on the Earth in an efficient manner. Text and tables are also commonly used to archive information about spatial objects. And as we discussed in Chapter 4, there are a variety of ways to use digital computer technology to store spatial data.

When thinking about spatial datasets, whether examining an existing dataset or setting specifications for developing future collections, there are a number of important non-spatial data elements to consider. The date of the data collection process is of course a natural item to record. It is one of the many important ancillary elements that provide us with an indication of the value of a dataset. There will be times when information from a specific point in time is needed. There will be other times when we need the most recent information available, as well as an understanding of how the landscape changes over time, to evaluate whether the data will be sufficiently precise or accurate for a specific application. Another consideration is the observation criteria and source. A vegetation map that is based on field visits of an expert botanist is of more value than a hypothetical map of presumed species distributions based on knowledge of latitude and weather. Thus, users must evaluate the characteristics of the best available, in order to determine their suitability. Other important elements to consider include the positional accuracy of the dataset, its logical consistency, and completeness.

Scale and resolution are two separate issues in spatial data. Consider two aerial photographs as a simple illustration of this fact. Both photographs were taken with the same camera, from the same altitude and location, over the same terrain. Assuming identical processing of the photographs, they will have the same scale. However, if one film emulsion has a finer grain and better contrast, it may have a better ability to resolve smaller details in the scene, and thus, better resolution. To put it in the terms we used in Chapter 4, we may say that the two images have essentially different minimum mapping units.

The geometrical properties of a set of spatial data are of course important. We must be able to determine the particular means used to specify location in the data. This usually implies that we must be able to identify the coordinate system and projection used in the data (as we discuss in section 6.6.1), or be able to devise a means to modify the spatial arrangement in the original data so that it corresponds to a desired arrangement (see section 6.6.4).

There are a number of other important attributes of a dataset, including information about accuracy and precision, as well as the density of observations used to develop the entire dataset. Regarding the latter, we may develop a large dataset based on a small number of measurements, and then use numerical models (as in section 6.7) to infer values at many other locations. This is commonly the case for elevation datasets, as well as detailed demographic information.

From a very simple point of view, we can distinguish two different families of datasets, and thus have an idea of the kinds of effort each requires before it is ready for use in a geographic information system. **Existing datasets** are those that are already compiled and available in some form. We of course do not minimize the fact that they may require a great deal of effort to make them appropriate for a particular use. The many steps required to prepare an existing dataset for use are discussed in Chapter 6. In contrast, there are many circumstances where we must **develop or generate the dataset ourselves.** In this second case, while we may have complete control over the data gathering process, we generally have much more work to do.

5.2 Existing Datasets

A lesson that many institutions learn through experience is that it is usually worthwhile to spend some time looking for existing data that can serve the stated needs, before plunging ahead and perhaps developing sources *de novo*. There is a great deal of spatial information available in the public domain for some parts of the world, if you know where to begin to look. We caution that different nations treat spatial data differently. In the United States, much spatial data collected by government agencies, including maps, photographs, and many kinds of digital data, are considered in the public domain. Thus, there are effectively no restrictions on access to this data. In contrast, in many other parts of the world, spatial data are considered the proprietary resource of the agency that collected the data, or may be controlled for economic or security reasons. In these cases, access to certain data may be strictly controlled. We will discuss some well-known sources of spatial data, both to help document these agencies and their data products, and to document the trend towards greater availability of digital cartographic data in the western countries.

The most common form of spatial data is a map. Section 2.1 briefly introduces cartographic products. Maps of various kinds are in common use for many kinds of spatial analysis. National agencies in many developed countries have systematic collections of map products at various scales, and programs for distributing and maintaining these resources. When appropriate maps are available, a digitizing process (described in Chapter 6) permits us to extract the information from the flat map and place it into a digital computer.

Table 5.1 Examples of Data in Digital Form Available from the United States Government.

DATA TYPE	DATA SOURCE
▪ Topography	
Digital Elevation Model	USGS/NMD
Digital Terrain Data	DMA
▪ Land Use and Land Cover	USGS/NMD
Ownership and Political Boundaries	USGS/NMD
Transportation	USGS/NMD, DOE
Hydrography	USGS/NMD
▪ Socioeconomic and Demographic Data	USCB
Census Tract Boundaries	
Demographic Data	
Socioeconomic Data	
▪ Soils	USDA/SCS
▪ Wetlands	USFWS
▪ Remotely Sensed Data	NASA, NOAA

Abbreviations used in the table:

DMA	Defense Mapping Agency
DOE	Department of Energy
NASA	National Aeronautics and Space Administration
NOAA	National Oceanic and Atmospheric Administration
USCB	U.S. Census Bureau
USDA/SCS	U.S. Department of Agriculture Soil Conservation Service
USFWS	U.S. Fish and Wildlife Service
USGS/NMD	U.S. Geological Survey National Mapping Division

However, many people are surprised to find that coverage of the world, in terms of scale, currency, and the themes of interest, is quite uneven.

When spatial data may be found in a digital form, there may be significant cost savings since the digitizing process is not required. In the United States, a number of kinds of thematic spatial data in digital form are being produced on a routine basis. While more and more states, regional authorities, counties, and cities are creating such datasets, the Federal Government is the best-known source for many GIS users. Table 5.1 presents a number of examples of digital datasets, produced on a routine basis, that are available (or are being made available) from the U.S. Government.

Agencies and commercial firms involved in remote sensing of the Earth have large data holdings, and may also be able to provide new data acquisitions. At the present time, two well-established commercial firms, Eosat and SPOT Image Corp., provide access to the Landsat and SPOT series of remote sensing systems. These are discussed in Chapter 10. Historical datasets are available, as well as orders for acquisitions in the future, on a fee-for-services basis. National agencies, such as the National Oceanic and Atmospheric Administration in the U.S., can provide spatial data from the operational weather satellite programs.

As we discussed in Chapter 2, there has been an explosion in the development and applications of geographic information systems since the 1960s. This explosion has been touched off by achievements in two areas: the incredible advances in computer science and technology, and the increasing availability of spatially referenced data in digital form. As we shall see in later chapters, data in digital form are generally easier to place into a modern geographic information system. Data in analog form, such as photographs and printed tables, require a painstaking and sometimes expensive conversion to digital form.

The majority of map makers think of aerial photos as *source materials* rather than *data*. They point out that, in showing objects and phenomena, maps (usually) have uniform symbols, and that maps treat both distance and directions in carefully controlled ways. Aerial photographs, as well as other forms of remotely sensed data, likewise show objects and phenomena, but they lack both the interpretation and geometric control. However, advances in sensor technology, as well as in processing and analysis techniques, are slowly changing the views of the cartographic community. We reemphasize that all operational mapping programs in the U.S. Federal Government is based on remotely sensed data of various kinds and photogrammetric techniques

For conventional aerial photography the standard format is a black-and-white or color photographic print on paper approximately 9 by 9 inches. For vertical aerial photography acquired from a mapping camera for a private individual, commercial firm, or government agency, the tilt should not exceed 3 degrees from vertical. Tilt is defined as the angle between the optical axis and a straight line through the center of the camera's lens. The scale of this photography is dependent upon both the focal length of the lens of the camera and the height from which the photo is acquired (see Chapter 10). Most scales used by agencies of U.S. federal government, for example, range from 1:20000 to 1:40000. While the most commonly used scales of conventional photography range from 1:4800 to 1:24000, a wider spread of scales, from perhaps 1:500 to 1:90000, are in common use. Scales in the range 1:500 to 1:2000 maybe employed for detailed urban planning and recreation-area management applications, while smaller scales will most often be used for more general resource analysis application (such as large-area forest stand maps or regional land-cover mapping and planning applications). A more detailed discussion of aerial photography is beyond the scope of this text.

Another product that deserves attention here are **orthophoto** maps. Orthophotos are produced by an instrument called an orthophoto scope. The orthophotoscope removes the relief displacement found in any photograph and creates an essentially planimetric product (see section 6.8). Orthophoto technology can be employed to produce a variety of different photographic products. Photogrammetrists are capable of producing orthophotos as photographic prints or half-tones. These can then be lithographed along with overprints of map grids, contour lines, place names, and other cartographic symbols.

The U.S. Geological Survey currently produces two types of photoimage products: the orthophotoquad and orthophoto map. The orthophotoquad is essentially a photographic base-representation of the standard USGS 7.5-Minute series topographic map. The orthophoto maps include contours and colors representing water bodies, wetlands, forests, and so on. These maps tend to be highly selective in terms of the features that are labeled and in coverage. The advantage of these products for an analyst is that they have properties of both maps and photographs. They may be used like a map, in that there is careful control of geometry and strict control of the use of symbols and graphics. At the same time, objects or phenomena may be interpreted from these orthophotomaps as in any other aerial photo. This combination makes these products ideal for many GIS applications.

In the United States, data from the Department of Commerce, National Aeronautics and Space Administration (NASA), National Oceanic and Atmospheric Administration (NOAA), and the Department of Interior through the United States Geological Survey (USGS) have been particularly important. Specific examples of data here include detailed census datasets at a variety of levels of spatial aggregation, remotely sensed data (see Chapter 10), as well as land-use, land-cover, and digital elevation data from the USGS files.

For sites in the United States, the U.S. Geological Survey, via the National Cartographic Information Centers and Public Information Offices, provides a wealth of information in digital form (we note that these will soon be known as Earth Science Information Offices). The **U.S. GeoData** tapes are one good example. Digital line graph data tapes are sold by 7.5- and 15-minute blocks, which correspond to USGS 7.5- and 15-minute topographic quadrangle maps - the standard map base for the United States. The U.S. GeoData digital tapes contain information in four thematic layers, which can be obtained separately: boundaries, transportation, hydrography, and the U.S. Public Land Survey System. All of these layers are coded in a vector format, based on a latitude-longitude coordinate system (as discussed briefly in Chapter 4). Another series of U.S. GeoData tapes is for land use and land cover. The land-use and land-cover tapes are available in either vector or raster cell format, based on a Universal Transverse Mercator coordinate system. Another kind of information available from the USGS is the Digital Elevation Model. These are regular raster datasets of elevation values.

There is a long-range effort in the U.S. government to create a National Digital Cartographic Database (Eric Anderson, pers. comm.). This is based on the work of an interagency coordinating committee, to set standards for the format and content of digital spatial data throughout the federal government. The layers to be included in this database include hypsography, hydrography, land surface cover, surface features excluding vegetation, boundaries, positional control, transportation, other man-made structures, and the Public Land Survey System.

Not all governments choose to make these kinds of data part of the public domain, as is the case in the United States. In Great Britain, for example, spatial datasets like those we have discussed are generally viewed as private holdings of the appropriate national agencies. As such, users may need to negotiate with the agency for access to these data.

The United Nations Environment Programme is a spatial data user as well as a producer. Through the newly established Global Resources Information Database, with existing centers at the UNEP offices in Geneva, Switzerland, and Nairobi, Kenya, efforts are under way to collect and disseminate important spatial datasets for the globe, as well as provide certain kinds of assistance in spatial data collection and processing to less-developed countries. Sample datasets in the archives now include range and endangered species distributions for parts of the world, as well as small scale global datasets of soils and vegetation. We believe this an important initiative, and a mechanism for the more effective management of important large-area datasets.

In many cases, existing datasets may not be quite suitable for the application. For example, in looking for data for a particular project, one may find a certain readily available map that has an appropriate scale and an appropriate set of thematic categories. The map may, however, be old enough to raise a significant concern about its value for the intended project. In cases such as this there are a variety of techniques one can use to update the old map -- thus avoiding the expenditure of time, effort, and money that is required to compile a new map from scratch. Photogrammetric tools such as transfer scopes and projectors permit us to merger the information of older maps with later aerial photographs (see section 6.8). Similar results may be obtained in some cases by using image processing technology (as outlined in section 10.3).

There are other times when an existing dataset doesn't quite suit the needs of an application. In some cases, the desired information may be obtained from these datasets through inference. For example, to help decide on a good location for a shop that sells and repairs bicycles, one may wish to obtain some suitable demographic data. But available demographic data may not provide directly the information desired about the bicycle-riding population in an area. If, on the other hand, available datasets (such as those which are created by the U.S. Census Bureau) do give information about the age distribution of the population as a function of location, plus information about the location of certain educational institutions, the needed information may be inferred (with the help of whatever reasoning, observation, and other information). Thus, in southern California, where college-aged people living within four miles of a college or university are likely to own bicycles, potential locations for a bicycle shop -- near student housing or between student housing and the educational institutions -- could be identified.

Overall, it is not always possible to find information about the quality of existing datasets. The expensive but conservative option is to attempt to verify the accuracy and precision, either quantitatively or qualitatively. Unfortunately, we often avoid these issues, and simply use what we believe to be the best available datasets for the task at hand. This latter choice runs a serious risk of letting us fool ourselves.

5.3 Developing your own Data

There will be times and circumstances in which it is necessary to develop your own datasets. Existing information resources may not be relevant to the problem, or perhaps they are not sufficiently current. There may also be datasets whose validity is unknown, thus forcing us to collect our own data to either test the existing datasets or to replace them. These occasions have a particular advantage, in that they give us the opportunity to design the data acquisition program to meet our needs exactly. The usual disadvantage of such a program is, unfortunately, the expense of designing, implementing, and managing the data-gathering task.

Constructing new datasets involves field work of many kinds. Maps of terrain or of the location of certain cultural features may need to be created. Details of plant and animal populations, such as noting the occurrences of different species or determining the age distributions within a population, may be required for an environmental report. Knowledge of the general trends in groundwater elevation in an area, as well as changes through the annual cycle of discharge and recharge, may be needed to site a water well or a waste disposal facility. Knowledge of the presence or absence of archaeological remains at predetermined locations may be needed before permits may be granted for construction projects. As we discussed in section 5.2, we must always look for both direct observations of the target and inferences that can be made from other datasets. **Sampling design** is one of the important elements of any data gathering plan, where decisions are made about how to gather the data of interest. We will discuss some of the key concepts of sampling design in the next few pages.

We must first take a moment and distinguish **accuracy** from **precision**. When making observations and measurements of any kind, an understanding of these two different concepts is very important. By accuracy, we mean freedom from error or bias, and thus, closeness to the "true" value. Precision, on the other hand, refers to our ability to distinguish small differences.

Determining the distance between two trees with a device that measures to the nearest centimeter provides a higher level of *precision* than using a device that measures to the nearest meter. On the other hand, if the centimeter-scale device is calibrated incorrectly, so that it is consistently underestimating distances, the meter-scale device may have higher *accuracy*.

Before looking into some of the details of sampling in space, we will examine some concepts of spatial pattern. In Figure 5.1, we show three different spatial patterns. These patterns might have come form an exercise in mapping the locations of certain types of plants. In Figure 5.1a, the pattern could be described as **clumped**, since the mapped objects are concentrated in certain areas. Another way of describing this distribution is to say we have **positive autocorrelation:** the objects are typically found close to others of the same kind. In the case of plants this might be due to the environment in these locations being particularly hospitable to the type of plant. In Figure 5.1b, on the other hand, the objects in the spatial pattern are spread out or **dispersed;** we could also say that we observe **negative autocorrelation**, meaning that the objects are typically found well away from others of the same kind. This type of pattern is sometimes seen in desert plants that compete for water, or in established forests where, in the competition for sunlight and root space, the

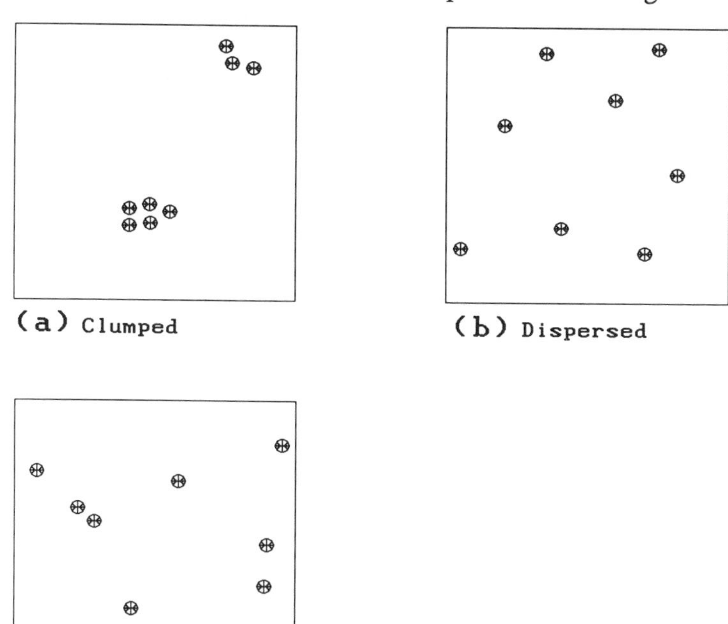

Figure 5.1 Spatial distributions. (a) clumped, (b) dispersed, (c) random.

older and more successful trees have overshadowed, crowded out, and killed off the smaller and less vigorous trees of like requirements.

One simple way to characterize these patterns is to focus on the distance from any given plant to the nearest neighboring plant of the same kind. In the clumped pattern, the average distance to the nearest neighbor is very small, and there will be relatively little variation in this distance if the clumps are uniformly dense. In the dispersed pattern, the average distance to the nearest neighbor is quite large. A **random** distribution of these plants would imply that at any location, there is an equally likely change of finding a plant, regardless of other plants in the vicinity. Thus, in a random pattern (Figure 5.1c), some plants would be close to their neighbors, and others would be far away. The clumped and dispersed spatial patterns in Figure 5.1 represent significant departures from random.

When designing a sampling program, one must make a number of decisions. One of the essential choices involves the samples themselves. **Point sampling** involves determining the desired information at a single point. For example, once a point is chosen, we could determine depth to groundwater by drilling a well at that location. Another option is **line sampling**, where two points are chosen and we census the desired information along that line. This is common practice in vegetation ecology, where all the plant species are identified along chosen lines or **transects**. The third common option is to locate appropriately sized areas or **quadrats** in the region of interest, and determine the needed information in each area. Quadrats are typically square or round. An advantage of quadrat and transect sampling techniques are that they provide information about spatial distributions both *within* the samples (the quadrats themselves) and *between* the samples.

The endpoints in a continuum of sampling strategies are when we take either a single measurement to infer regional characteristics, or when we have **exhaustive sampling**, and is the case where we are able to sample the entire region. In essence, the latter is the extreme case of quadrat sampling, in which the area of the quadrat is chosen to be identical to the area of interest. Clearly, this will provide us with the best information, since we will examine the geographic region of interest completely. However, we often have limited resources with which to conduct our data gathering, and thus have to design a plan to maximize the value of the limited data we can obtain.

Figure 5.2 shows two different ways to apportion sampling effort. In the upper left of Figure 5.2, we show an example of **random sampling**. We

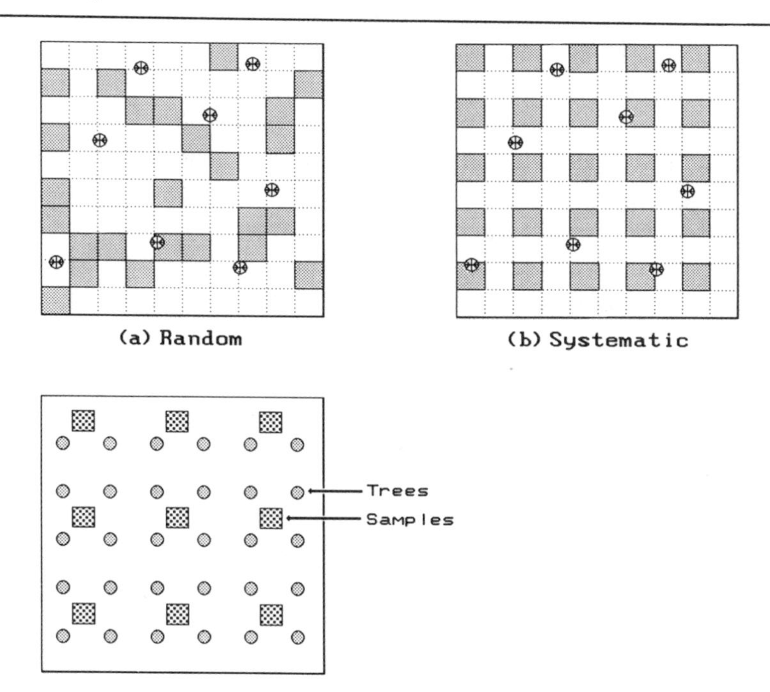

Figure 5.2 Sampling strategies, random and systematic.

have first divided the area into quadrats of equal size, and then, using a computer-generated list of random numbers, have randomly selected 25% of the quadrats for examination. Note that this is but one way to select the transects. Alternatively, we could have chosen the quadrat center points randomly, rather than choosing from a regular array of possible quadrats. In this example, our 25% sample has discovered 1 object of interest, and we thus extrapolate to decide that 4 objects probably exist in the region. Notice that the true number of objects is 8; we have underestimated the total.

Another way to apportion our sampling effort is shown in the upper right of Figure 5.2. In this second case we have used a **systematic sample**. In this case, we use the same number of sample quadrats, but they are arrayed in a regular, predictable pattern in space. We find two objects in this second sample, and thus extrapolate to a total population of 8.

The example above is not meant to suggest that systematic sampling is always better than random sampling. Systematic sampling is often easier to design and implement than random sampling, since it is a relatively simple task to systematically place the sampling elements, whether they are points, lines, or quadrats. Random sampling, on the other hand, requires relatively more effort to choose the locations, navigate in space, and locate the samples. On the other hand, systematic sampling can cause all kinds of artifacts when the sample spacing roughly coincides with a multiple of the spacing of the objects of interest. Consider, for example, trees in an orchard that are routinely planted in a square array, with 4 meters between rows and columns. If we apply quadrat samples systematically on a square grid with an 8-meter spacing, precisely twice the spacing of the trees, we could completely miss all the trees (Figure 5.2c)! To deal with this problem, a method called systematic unaligned sampling is used, in which the sampling interval and orientation is specifically adjusted to avoid alignment with the phenomena of interest (Congalton, 1988).

Another sampling strategy is used when we have some information about the study site. This third option, known as **stratified sampling**, involves dividing the study region into relatively homogeneous units, and sampling each unit separately. For example, imagine a program for monitoring a group of insect pests that damage farm crops. The target area may have several different crops under cultivation. In a stratified sampling program, we can explicitly take the spatial distributions of the crops into account. We might sample more intensively in fields growing crops that the target insects are known to inflict great economic damage upon, to be able to take measures as early as possible in the growing season. In contrast, we might work less hard in crops on which the insects are known to cause little damage.

An important use of stratified sampling is seen when the geographic region has important but rare polygonal areas. Consider a program to determine the distribution of land cover as a function of cover type. In a random sample, the number of observations of a particular type of land cover will be proportional to the area of the cover. Thus, common land-cover types will be sampled frequently, and rare land-cover types may never be observed. To compensate, we can use a stratified sampling plan to make sure that we visit examples of every known land-cover type, and thus improve our knowledge of the rare types.

It is common practice in many fields to use a **pilot study** -- a rapid or preliminary look at the population of interest -- before any major sampling effort. The pilot study is usually designed for two purposes. First, it permits us to gather some information in the field, possibly from the target area. This small amount of information can be used for adjusting quadrat size, selecting the total number of samples for the principal sampling effort, testing the observational methods, and (where possible) developing some general characteristics of the study population. Such a pilot study could permit us to choose an unaligned sampling strategy in the case illustrated in Figure 5.2c, and thus could prevent us from being misled. In the case of stratified sampling, the pilot study could provide the necessary information to develop or test the initial stratification. The second purpose of the pilot study is that it permits us to check our original hypotheses about the costs and time that would be required for gathering the data. Thus, the pilot study permits us to avoid costly mistakes and poor data quality.

The tools of remote sensing, discussed in some detail in section 6.8 and Chapter 10, can be of tremendous value in designing field studies. Remote sensing, defined here as a suite of techniques for making observations at a distance, can often provide cost-effective information about properties of the Earth's surface over large areas. Frequently, an aerial photograph or processed multispectral scanner image can provide the basis for a field sampling campaign. In our own work, these data sources have been used to minimize the effort required to collect more traditional spatial datasets, such as social and environmental surveys.

Chapter 6

Preprocessing

Preprocessing procedures are used to convert a dataset into a form suitable for permanent storage within the GIS database. Often, a large proportion of the data entered into a GIS requires some kind of processing and manipulation in order to make it conform to a data type, georeferencing system, and data structure that is compatible with the system. The end result of the preprocessing phase is a coordinated set of thematic data layers.

The essential preprocessing procedures include:

- format conversion;
- data reduction and generalization;
- error detection and editing;
- merging of points into lines and of points and lines into polygons, where appropriate;
- edge matching;
- rectification/registration;
- interpolation; and
- photointerpretation.

We'll examine each of these in turn in this chapter. At the same time, many of these procedures are valuable at other stages in the end-to-end spatial analysis problem. We will point out these additional uses as they come up in later chapters.

6.1 Format Conversion

Format conversion covers many different problems, but can be discussed in terms of two families: conversion between different digital data structures and conversion between different data media. The former is the problem of modifying one data structure (such as those discussed in Chapter 4) into another. The latter typically involves converting source material such as paper maps, photographic prints, and printed tables into a useful computer-compatible form. Note that the reverse of data medium conversion -- changing the data stored within the digital database into maps, prints and tables -- is the problem of generating output products, which we discuss in Chapter 9.

6.1.1 Data Structure Conversions

In Chapter 4, we discussed several different data structures. There are many times when different datasets gathered for the same project are expressed in different data structures. Part of the cause of this lack of homogeneity is the nature of the datasets themselves, some data structures being more suitable to some kinds of data than others. This problem is becoming more acute as increasing amounts of current data are created and maintained in various digital forms, while historical records are almost universally stored on paper or film. Another kind of problem that arises frequently is when we have raster datasets, such as digitized photography and multispectral scanner data (discussed in Chapter 10), and our GIS is based on a vector data structure. In these cases, we must be able to inter-convert between the data structures. Similar conversions will be required in order to develop final output products as well (see Chapter 9).

The simplest forms of conversion are between members of a family of structures. For example, there are several common raster formats for raster data. Typical raster GIS datasets include arrays of elevation, rainfall, and classes of land cover. This is also the kind of data that is produced by multispectral scanners, which are common sensors on both aircraft and spacecraft platforms (some applications of these systems are discussed in Chapters 10 and 12). The data produced by such systems may be thought of as an array of brightness values for each wavelength band in the sensor. These systems generate datasets that are comparable to any other multivariate collection of raster data, including problems of geometric registration between the wavelength bands or between different dates of acquisition (Welch, 1985).

There are, however, several ways to organize such datasets. Keeping each variable's data (for example, elevation and annual rainfall, or the different spectral channels from a multispectral scanner) as a separate array is one common method. This method is often called **band sequential** (BSQ), since each array is kept as a separate file on the magnetic disk or tape. In this case, one data file would contain the elevation array, and a separate file would contain rainfall values. A common alternative, called **band interleaved by pixel** (BIP), places all of the different measurements from a single pixel together. This organization may be thought of as a single array containing multivariate pixels (Figure 6.1). The first element in this second format would contain the elevation value for the pixel in the first row and column; the second element would contain the rainfall value for the same pixel.

When operations on the data involve a single theme or layer at a time, a BSQ raster database can be the most efficient organization. This is because the specific theme of interest (or spectral channel) can be analyzed and manipulated as a physically independent entity. Conversely, when working with more than one data theme at a time, the BIP organization can be the most efficient. For example, consider a raster dataset with two themes: elevation data points and classes of forest cover. If a principal activity is to operate on a single data layer at a time, such as deriving slope from the elevation data points, the BSQ organization makes the elevation data directly available, without having to read the data files past land-use values. If, on the other hand, an analytic operation requires comparison of both themes on a pixel-by-pixel basis, such as finding the location of the highest elevation for each of several forest-type classes, the BIP organization makes good sense, as the values of the two themes for each pixel are adjacent in the database.

The **band interleaved by line** (BIL) raster organization is a middle-ground between the extremes of BSQ and BIP. In this form, adjacent ground locations (in the row direction) for a single theme are adjacent in the data file, and subsequent themes are then recorded in sequence for that same line. In this way, the different themes corresponding to a row in the file are relatively near each other in the file. Thus, one expects that its performance on specified tasks will fall between the pure sequential or pure pixel-interleaved forms. This intermediate type of multivariate raster is used in some commercial raster systems.

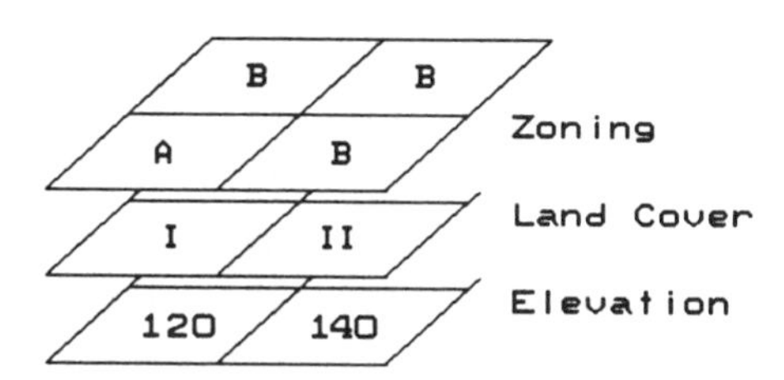

1. Band Sequential - BSQ

File 1. Zoning: A, B, ...

File 2. Land Cover: I, II, ...

File 3. Elevation: 120, 140, ...

2. Band Interleaved by Pixel - BIP

Line 1:

A, I, 120, B, II, 140

pixel 1 pixel 2

3. Band Interleaved by Line - BIL

A, B, ...
I, II, ...
120, 140, ... Line 1
B, B, ...
... Line 2

Figure 6.1 Common raster data organizations. In band sequential data, each variable is stored in its own file. In band interleaved by pixel data, all the information about a pixel is kept together. In band interleaved by line data, all the values of a variable from a single line are stored before the values for another variable in the same line.

For those readers with a data processing background, we will briefly add to the complexity. There are two common physical data organizations for BIL-structured data. In one, the physical records hold all the themes from a single row in the array, ordered as described in the previous paragraph. Thus, the number of physical records is the same as the number of rows in the array. In the other common BIL format, a physical record corresponds to a single thematic category. Thus in this second case, the number of physical records in the dataset is the product of the number of rows in the array times the number of unique themes.

The problem of converting between the different raster data formats described above is relatively simple. Typically, a portion of the data in one format is read into a memory-based storage array, and the appropriate pointers created to extract the data values in whatever sequence is required for the new format. Optimizing such conversion software on a given computer is straightforward. We discuss another family of data format complications in Chapter 10.

There are a wide variety of problems that develop when converting datasets between different vector data structures. As we noted in Chapter 4, there are many different vector organizations. As a guiding principle, it is expensive to generate topological information when it is not explicitly present in the vector data structure. To illustrate this point, converting data from an arc-node organization to a relational one is very easy. As we observed in the last chapter, these are very similar data structures, in terms of the way topological information about spatial objects are organized. In effect, these two data structures are really storing the exact same semantic information, with a slightly different syntax.

At the other end of the spectrum, consider the problem of converting data in a whole polygon structure to an arc-node structure. In a dataset stored in whole polygon structure, there is very little explicitly identified topology. The list of nodes that form the boundaries of each individual polygon is stored. Consider just the problem of extracting the arc-node node list. We must go through the entire list of polygons, and create a list of the unique nodes. This might require a double-sort of all the points in the polygon file, and then a pass through the sorted list to identify the unique nodes. Creating the arc list requires another pass through the whole polygon file, this time cross-referencing edges of each polygon to the corresponding elements in the node list and generating the appropriate pointers. Furthermore, to identify all the polygons that border a given vector requires another complex sorting operation to identify all the shared edges.

Converting vector data into a raster data structure is conceptually straightforward, although practically difficult. For point data elements, the cell or pixel in the raster array whose center is closest to the geographic coordinate of the point is coded with the attribute of the point. Thus, the elevation value from a surveyed benchmark is transferred to the raster cell whose location is closest to that of the original point. Of course, this operation usually changes the stored location of the point -- it is unlikely that the original point location exactly coincides with the center of a raster cell. This approach also ignores the problem of different objects occupying the same cell. Because of these important limitations, the conversion from vector to a raster data structure is not normally reversible: we cannot retrieve the original data points from the derived raster data without error. For some operations, this can be a fatal flaw.

For linear data elements, the data structure conversion can be visualized by overlaying the vector or linear element on the raster array (see

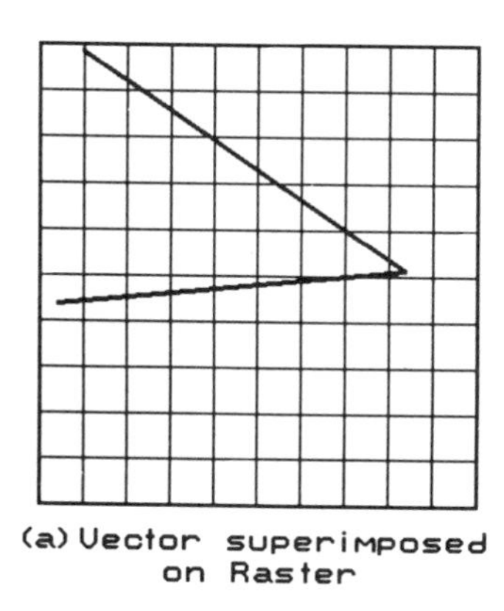

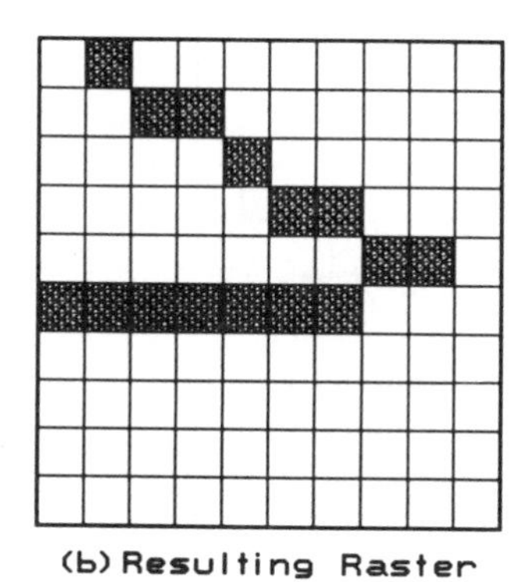

Figure 6.2 Vector to raster conversion.

Figure 6.2a). The simplest conversion strategy would be to identify those raster elements that are crossed by the line, and to then code these cells with the attribute or class value associated with the line. For lines that are not oriented along the rows or columns of the array, the raster representation shows a stair-step distortion (see Figure 6.2, as well as the discussion of aliasing in Chapter 9). In this first discussion we have ignored the problem of specifying the thickness or width of rasterized line (Peuquet, 1981b).

Polygons can be converted to a raster structure in two steps. First, the line segments that form the boundaries of the polygon can be converted as in the last example, producing what is sometimes called the **skeleton** or **hull** of the polygon. Second, those raster elements contained by the polygonal boundaries are recoded to the appropriate attribute value (Figure 6.2b).

Converting raster data to a vector data structure can be a great deal more complex. To keep the discussion brief, we will first examine data where object boundaries are a single pixel wide, though this is rarely the case in real life. A simplistic approach for a trivial binary image (where the data are either in class 0 or class 1 and lines are only a single pixel wide) is illustrated in Figure 6.3. Consider that each raster cell can be represented by a point in its center. We can then draw vectors between all the non-zero 4-connected raster elements. Such an algorithm models all possible vectors as orthogonal vectors, parallel to the rows or columns of the raster. Further, this algorithm constrains all vectors to be of discrete lengths that are equal to an integer multiple of the raster spacing. Of course, vectors in real life do not have either of these restrictions.

This approach will not be able to recover all vectors that have been

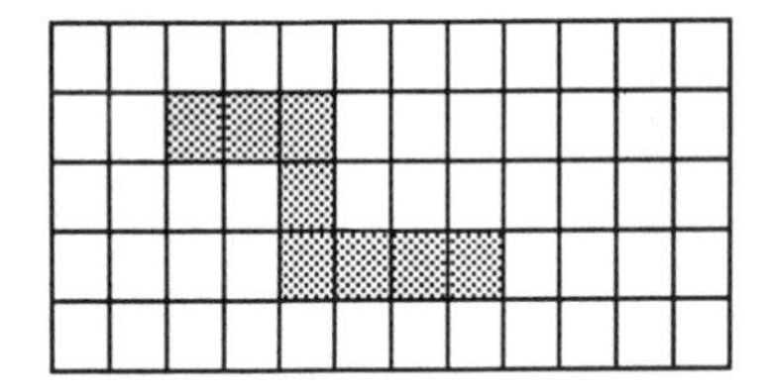

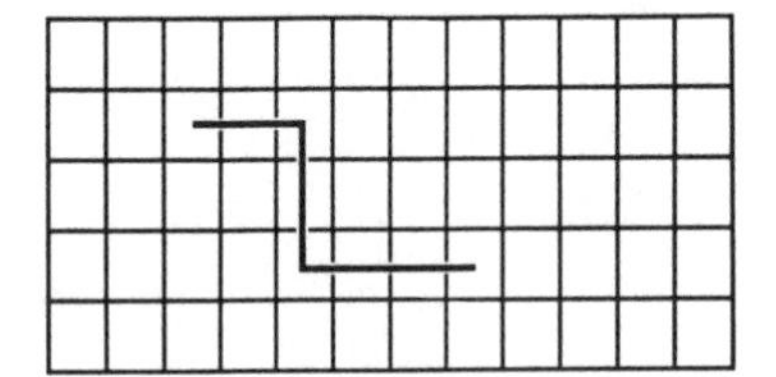

Figure 6.3 Simple raster to vector conversion.

converted to a raster. Consider the raster-coded vector in Figure 6.2a. Since the straight-line nature of this data element has been lost in the process of conversion, we will not be able to recover the straight line without ancillary information. However, there are algorithms that can be used to extract straight lines from raster data sets under restrictive circumstances, at an increased computational cost.

To illustrate a somewhat more sophisticated approach to raster/vector conversion, let's start with data that might have come from a flatbed digitizing tablet. The following table describes the (x-y) coordinates of three graphic objects, which we will convert to a raster, and then attempt to convert back to vectors. By returning to a vector representation, we can begin to develop an understanding of the limitations of the algorithms.

Object #1 -- a vector, defined by connecting (x-y) points:

1,62
12,62
12,51
1,51

Object #2 -- a vector, connecting (x-y) points:

16,57
20,60
27,54
34,57

31,48
22,52
15,52

Object #3 -- a complex closed region, bounded by the closed (x-y) path:
8,46
21,46
11,38
15,48
20,38

These three objects -- two simple open paths and a complex closed region -- are plotted in Figure 6.4a. We'll explore this data by starting with these three vector objects, and converting them to a raster representation as in Figure 6.2. Converting these intermediate raster data back to a vector form, based on the 4-connected neighborhood approach described in Figure 6.3, gives us Figure 6.4b. Notice that the upper left object in the figures is perfectly reconstructed in a 4-connected algorithm, but the other two objects are badly distorted. In particular, elemental line segments between adjacent raster cells that are **not** parallel to the raster axes have disappeared.

A more complete (but still simple) approach would be to search an 8-connected neighborhood around every raster element, searching for possible vector connections. In addition to the unit vectors orthogonal to the raster axes, this new algorithm permits diagonal vectors to be found. The output of this algorithm is shown in Figure 6.4c. The general characteristics of all three objects are retained. In this case, all the connections between the original points are preserved. However, this algorithm finds many connections between raster cell elements that were not in the original vector dataset.

A further improvement, involving significantly more computation, would be to draw only diagonal lines between elements in any 3-by-3 8-connected region when they are not already connected by 4-connected orthogonal vectors. This additional rule gives us the objects in Figure 6.4d. Where two lines derived from the rasterized boundaries of the objects are close, this algorithm draws an extra line, compared to the original (Figure 6.4a). However, this is a relatively successful recovery of the original objects.

Beyond these theoretical models, practical datasets require several additional functions. As Peuquet (1981a) explains, operational systems require two general functions for converting a raster to a vector dataset. The

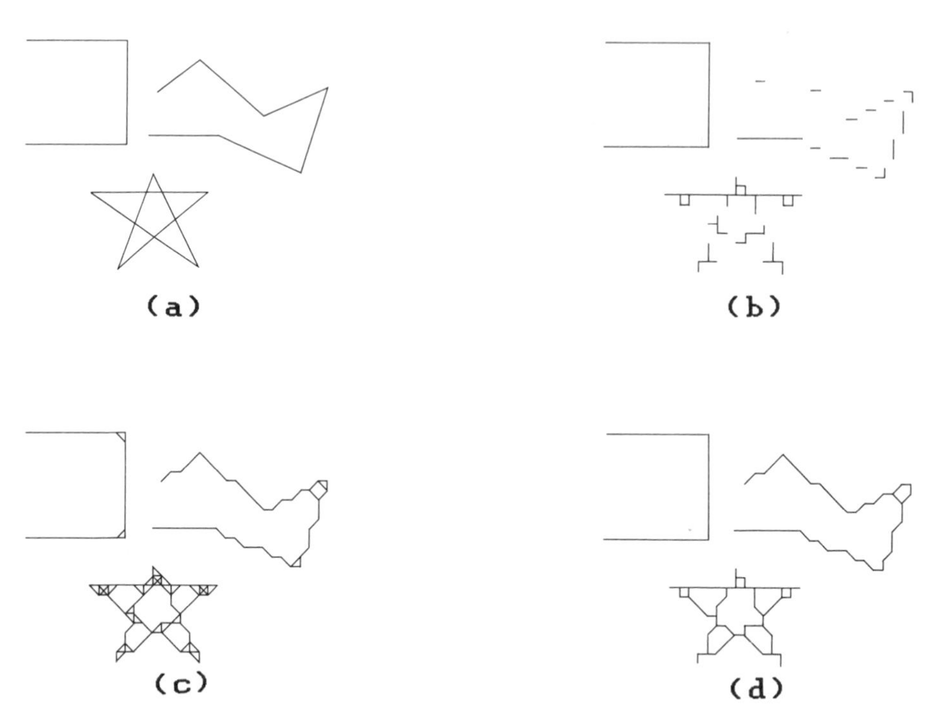

Figure 6.4 Complex approaches to raster-vector conversion. (a) data input, (b) 4-connected vector reconstruction, (c) 8-connected vector reconstruction, (d) 8-connected reconstruction with redundancy elimination.

first, **skeletonizing** or thinning (Figure 6.5), is required because the input data are not generally as simple as those we have discussed: single pixels for point data, and unit-width vectors for both vectors and polygon boundaries. Algorithms for determining the skeleton of an object are sometimes described as a peeling process, where the outside edges of thick lines are "peeled" away, ultimately leaving a unit-width vector. A symmetrical alternative approach is to expand the areas between lines, with the same ultimate goal. In either of these cases, the process is a sequence of passes through the data, with each pass producing narrower vector outlines. A third alternative, the medial axis approach, is designed to directly identify the center of a line, by finding the set of interior pixels that are farthest from the outside edges of the original line.

After the raster data have had such thinning operations applied, the vectors implicitly stored in the raster are extracted. The extraction process may be based on the models discussed above. Finally, the topological

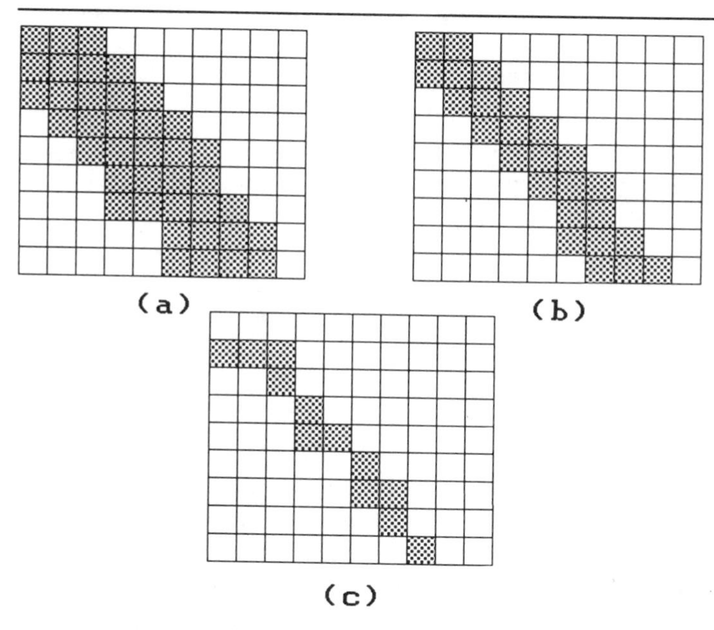

Figure 6.5 Skeletonizing. (a) original dataset, (b) data after one thinning pass, (c) final skeleton.

structure of the lines is determined, by recognizing line junctions and assembling the separate segments into connected vectors and polygons.

Crapper (1984) presented an interesting analysis of the relations between vector and raster thematic data. His concern was understanding the accuracy of thematic maps. Consider an original dataset that is a map or photograph. Overlaying a grid on the original data and making assignments of thematic category, is relatively simple for cells in the interior of a homogeneous polygon. The difficulty arises at the boundaries between polygons, where single grid cells cover more than one category. As we have mentioned, we could use a plurality rule to assign classes to the boundary cells, or alternatively, we could label them as a unique class. Crapper derives a relationship to estimate the number of boundary cells in this process, and thus gives us insight into the total area of boundary cells.

6.1.2 Data Medium Conversion

Most of the spatial data available today are not in computer-compatible formats. These include maps of many kinds and scales, printed manuscripts, and imagery (based on photographic processes, or generated by non-photographic instruments). Converting these materials into a format compatible with a digital geographic information system can be very expensive and time-consuming. According to the U.S. Geological Survey's Technology Exchange Working Group (U.S. Department of the Interior, 1985, p. iii):

> *The digitizing of conventional cartographic data is perhaps the most resource intensive phase of constructing a digital cartographic data base or utilizing a geographic information system.*

The most common means of converting maps and other graphic data to a digital format is to use a **digitizing tablet** (Figure 6.6a). A digitizing tablet system consists of several parts, among which is a flat surface on which the map or graphic to be digitized is placed. This flat surface is typically from 1 to 20 square feet in area, and may be back-lit or even transparent to permit digitizing from transparencies. The user traces the features of interest with either a pen-like stylus or a flat cursor. The electronics in the tablet system convert the position of the stylus or cursor to a computer-compatible digital signal, with a typical precision of 100 to 1000 points per inch.

There are several technologies used in commercial digitizing systems. Acoustic systems are relatively low in cost, and often able to work with large-format materials. Such systems use acoustic transducers to triangulate positions on the map or graphic. Electromagnetic and electrostatic systems are also available, and are generally preferred when high accuracy and precision are required. When the tablet has a cursor for tracing data elements, there are often buttons or switches on the cursor itself. This is particularly helpful, since it permits the analyst to select functions without moving from the map or graphic to a computer keyboard.

When it is necessary to digitize a map or other graphic with very high precision, the dimensional stability of the medium can become important. Photographic films are considered very stable with respect to changes in temperature and humidity, with distortions below 0.2 percent (Wolfe, 1983). In contrast, paper shrinkage or expansion can range up to 3 percent, depending on paper type and thickness, as well as processing methods. These may be important considerations.

When beginning a session with a digitizing tablet, the user must specify a number of attributes of the map, as well as the map's location on the digitizing tablet. Typically, the user will be prompted by the system for information about a map's scale and projection; menus with common choices can help the user to enter this information quickly and accurately (see Figure

(a)

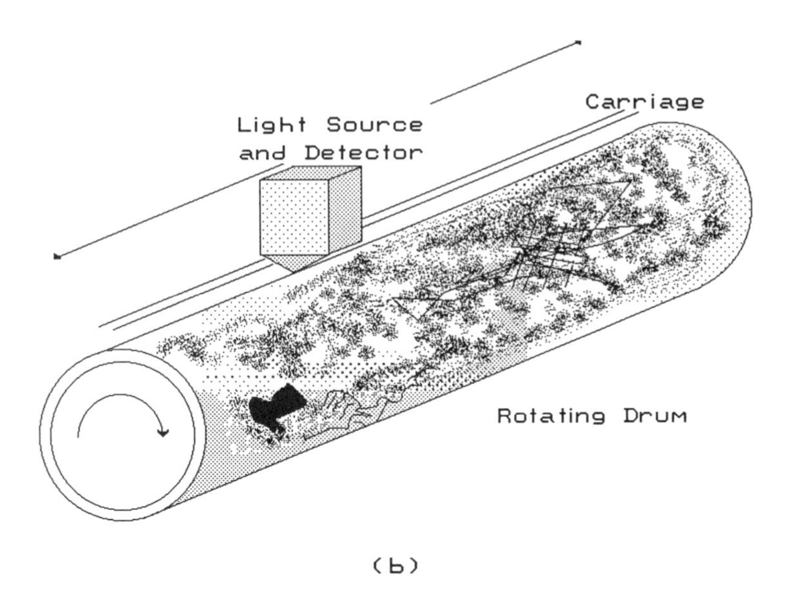

(b)

Figure 6.6 Conversion tools. (a) digitizing tablet, (b) scanning system.

6.7). After entering this information, the tablet or stylus is used to specify both georeferencing information (for example, by placing the cursor at locations of known latitude/longitude) and a region of interest. For well-known map projections on most systems, these procedures permit any subsequent location of the stylus or cursor to be converted unambiguously into a geodetic location.

One of the functions a user should be able to select is the mode of digitizing. In **point mode**, individual locations on the map (such as elevation benchmarks, road intersections, or water wells) can be entered by placing the cursor over the relevant location and pressing a button. In **line mode**, straight line segments (such as short segments along political boundaries, straight road sections, or lines of constant bearing on appropriate map projections) are entered by moving the cursor to one end of the line, pressing a button on the cursor, then moving to the other end and pressing a button again. The system automatically converts these two entered points into an appropriate vector. Digitizing curved line segments in this manner can be very exacting work. In **stream mode**, the location of the cursor on the map surface is determined automatically at equal intervals of time, or after a specified displacement of the cursor (so that points are approximately evenly spaced; Nagy and Wagle, 1979). Stream mode is particularly useful when digitizing curved line segments, such as the boundaries of waterways. However, in stream mode it is often too easy to create very large data files, since data points are entered into the system so quickly. Further, stream mode can be very demanding on the operator.

Neglecting the question of accuracy, digitizing tablets have finite resolution. We normally consider these devices as operating fundamentally for vector input, since we can ostensibly locate any point on the surface of the tablet. However, the finite limits on the precision of these devices provide an underlying raster-like limiting resolution element.

Scan digitizing systems, generally called optical scanners or scanning densitometers, are typically larger and more expensive than digitizing tablets (Figure 6.6b). With many high-precision systems, the map (or graphic) to be digitized is attached to a drum. A light source and an optical detector are focused on the surface of the drum. The drum rotates at a constant speed around its major axis. Because of the rotation of the drum, the detector traces a line across the map, and the electronics in the associated computer system record numerical values that represent brightness (or color) on the map. This traced line across the map corresponds to a single row in a raster of data values. The detector then steps along the axis of the drum, in effect moving

```
   DIGPOL -- Digitize Polygons, Vectors, Points      Version 7.2.06.67
      Copyright (C) 1986 ERDAS, Inc.  All rights reserved.
      Installation : U. of California (Santa Barbara) Remote Sensing Unit
-------------------------------------------------------------------------
Enter Output filename: test1

Setup NEW map or use PREVIOUS setup? (N,P) [New] :

Select type of coordinates to use:
  -  UTM                            -  Longitude / Latitude
  -  State Plane                    -  Other
 (U,S,L,O) [UTM] : UTM
Select SCALE of this map:

1)    1:24,000    (1"=2000 Feet)   5)    1:63,360    (1"=1 Mile)
2)    1:25,000                     6)    1:100,000
3)    1:50,000                     7)    1:125,000
4)    1:62,500                     8)    1:250,000  (1"=Approx 4 Miles)
                                   9)    Other (i.e. 1: _____)
      ? 4

Enter UTM X of Reference Point ? 315360
Enter UTM Y of Reference Point ? 1625475

          Digitize LEFT    Reference Coordinate

          Digitize RIGHT   Reference Coordinate

      *   Digitize BOTTOM Reference Coordinate
-------------------------------------------------------------------------
 Number of Digitized Points= 11
 Data Value = 10
 POLYGON Mode
X,Y=   317141.25   1624303.25


       Use "A","B" or "C" on Keypad
       to indicate new data value.

       Polygon Mode= AnnnA
       Vector= BnnnB   Point= CnnnC

 *     Digitize points by pressing
       button "1". Back up by pressing
       the "E" button. After the last
       point, press button "2".

       If "A","B" or "C" not entered,
       the previous data value is used.
       Continue digitizing polygons
       using buttons "1" and "2".

       Signal end of job with key "D".
```

Figure 6.7 A digitizing menu. (Courtesy ERDAS, Inc. Copyright 1986. Used by permission.) Setting up a new session requires specifying the map coordinate system (UTM in this case), scale (1:62500), and then indicating a control point coordinate location. Left, right, and bottom reference coordinates then are indicated to calibrate the system for any relative rotation between the digitizer surface and the map's coordinate axes. Points, lines, and polygons are then digitized, with the user controlling the system from the 16-button cursor on the tablet.

the detector to a new row in the raster, and the process repeats. In this way, the original map is converted to a raster of brightness values.

An alternate mechanism uses a line array of photodetectors, which sweeps across the map in a direction perpendicular to the array axis (comparable to the system illustrated in Figure 10.4). Such a device has a much simpler mechanical design than the drum systems mentioned above. In modern systems of either type, the map or graphic to be scanned can be on the order of one meter square, and the scanning step size as small as 20 micrometers. Based on these extreme values, the resulting raster dataset for a one-meter-square map could contain 2.5×10^9 pixels, before additional operations are begun. The resulting scanned raster of brightness (or color) may be used directly, or software can convert this initial raster dataset into other forms.

A scanning system may be sensitive to 100 or more shades of brightness or color in the source document. These shades of brightness, along with information about spatial patterns in the raster, can be processed by the appropriate software so that the various graphic objects in the source documents can be distinguished. When required, the system may convert the raster dataset to a specified vector format (as discussed in section 6.1.1) as well as compress the dataset size in various ways.

A normal sequence of steps in the use of a scanning digitizer would start with the actual scanning process. According to one set of figures, scanning a 24-by-36 inch document at 20 lines per millimeter takes 90 minutes. The raster files are then interactively edited, to ensure that the skeletonizing process accurately extracts the graphic elements in the original dataset. Next, the preprocessed raster data are converted to a vector dataset. The vector data are then structured, to build whatever composite elements (e.g., chains as connected line segments) and topological relations (e.g., containment and adjacency) are required in the ultimate datasets. Finally, the data are interactively edited for quality assurance requirements, and any additional attributes entered and verified.

In many cases, there is tremendous redundancy in the raster digital data developed by an automated scanner. One way to store the data with less redundancy is to use **run-length encoding**, which we discussed in Chapter 4. in For example, if we were to scan a page from this book, there would be a long sequence of identical white pixels from the row at the extreme top of the page, since there is no text at the top. Rather than placing many such sequences of identical pixel values in the data file, it is more efficient to use a run-length encoded file, where we store a single copy of the brightness value and the number of repeated pixels of this value.

Map-like images are often successfully compressed in this way, due to two characteristics of maps: (1) there are generally few unique brightness (or color) values, and (2) there are often horizontally homogeneous sections. In contrast, this strategy rarely works as efficiently with photograph-like images, since the broader dynamic range and generally high texture of these data types provide a smaller opportunity for many horizontally homogeneous runs. A side benefit of a run-length encoded image is that this data structure explicitly codes for the boundaries in a raster array.

Peuquet (1981a) presents case studies of several systems that have been used operationally for converting data from conventional maps to digital datasets. In these systems, the overall conversion process includes scanning the maps, which converts the continuous maps into a discrete raster, extracting the line segments and developing the topologically structured dataset.

In section 6.8, we briefly mention more sophisticated means of extracting three-dimensional data from groups of photographs. These techniques are the domain of photogrammetry, and beyond the scope of this text.

6.2 Data Reduction and Generalization

An input record may require data reduction of various kinds. For example, field crew records of tree-crown diameter may contain several measurements at different points around the circumference of the tree. In a given application, we may need to average the measurements of a single tree, and enter the average into our database. As another example, existing datasets may record vegetation by species, and our requirements may be satisfied by simply recording by species groupings (i.e., deciduous versus coniferous trees).

A more complex form of data reduction involves changes of scale in spatial data. We may need to assemble property ownership records in an area, and the original surveyor's records may be more detailed than we require. There are two obvious options: either accept the level of detail (and thus, incorporate a greater volume of data than necessary, with the attendant increased processing and storage costs), or develop a less precise representation from the original source data. The latter is called

generalization (Figure 6.8). For vector data, consider a locus of points, connected by straight line segments, which describe a coastline. In this case, a naive generalization procedure could involve simply replacing pairs of neighboring points with their average geographic coordinates. In this way, we reduce the total number of points by a factor of two, and in the process, reduce the level of detail in the dataset.

A more sophisticated approach would involve modeling the behavior of a modest number of successive points by a numerical algorithm. For example, a polynomial curve of specified order can be least-squares fitted to a sequence of data points, and then a smaller number of points along the path of the fitted polynomial recorded in their place. Another alternative is to eliminate points from the original dataset that are too close together for a specified purpose, or fall along a straight line (within a specified tolerance). The former procedure

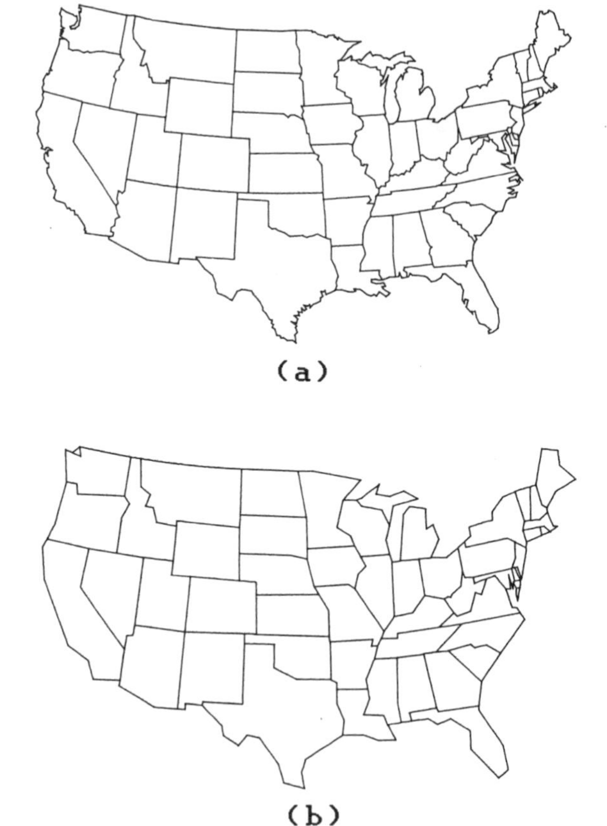

Figure 6.8 Generalization. (a) original vector data, (b) generalized vector data.

is a common operation in many computer graphics systems, when points in the original dataset are too close to resolve on the desired graphics device. In a plotter, for example, the limiting resolution could be a simple function of the width of a pen and the repeatability of the plotter's positioning machinery. By repeating these processes, depending on the dataset and the desired resolution, we can derive a lower-resolution representation of the original vector dataset (Figure 6.8b). This approach is often termed **thinning**.

For continuous raster data (such as millimeters of rainfall per year, or elevation), we can similarly create a more generalized dataset in two steps. First, compute the average value of the attribute in a two-by-two neighborhood. Second, record this average value in a new raster cell at the geographic location of the point shared by the four original raster cells. This kind of procedure, called **resampling**, can also be used when the required new raster cells are not of a length that is an integer multiple of the initial cell length. This is directly related to the problem of rectification discussed in section 6.6; see Moik (1980) for further details. For nominal and ordinal raster data, rules for aggregation must be developed. Common aggregation rules include determining the majority or plurality class in the averaging neighborhood, as well as defining hierarchies of classes (for example, aggregating records of individual plant species at one raster resolution into species groupings at a coarser resolution).

6.3 Error Detection and Editing

In any information system, facilities must be provided to detect and correct errors in the database. Different kinds of errors are common in different data sources. To illustrate some of the common varieties, we will discuss some of the errors that must be detected when generating a vector dataset, whether developed from a manual session with a digitizing tablet or an automated scanner.

Figure 6.9 illustrates some common errors of the digitizing process. Polygonal areas, by definition, have closed boundaries. If a graphic object has been encoded as a polygon (rather than a vector or point), the boundary must be continuous. Software should be able to detect that a polygon is not closed. Causes for this kind of error included encoding error (the object is a vector, rather than a polygon) and digitizing error (either points along the boundary of the polygon, or connections between the points, are missing).

With some systems, the boundaries between adjacent polygons may have to be digitized twice -- once for each polygon on either side of the shared vector. This also happens on more advanced systems when the document cannot be digitized in a single session. Such systems create **slivers** and **gaps**. Slivers occur when the adjacent polygons apparently overlap; gaps occur when there is apparent space between adjacent polygons. Similar problems occur when several maps are required to cover the area of interest. In this case, spatial objects may be recorded twice because of overlapping regions of coverage. An arc-node database structure avoids this problem, since points and lines may be recorded once and then referenced by many different spatial objects.

Errors of several kinds can appear in the polygon attributes. An attribute may be missing or may be inappropriate in a specified context. The latter kind of error is often related to problems in coding and transcription, such as recording an elevation value into a land-use data layer. Software to detect such errors is vital, and it can be very straightforward to develop and use.

In the practical use of a geographic information system, where there is a great deal of attribute information to include in the database, an interactive capability of correcting errors is vital. A development effort at one of the U.S.

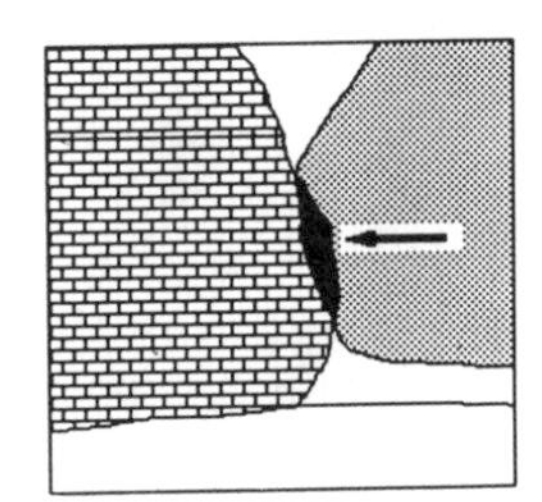

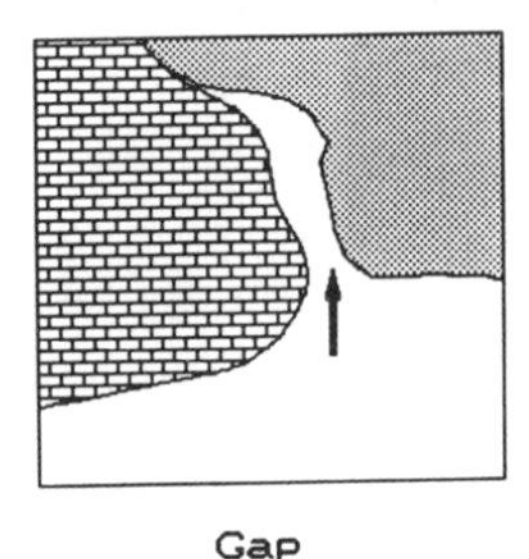

Figure 6.9 Digitizing errors.

federal agencies several years ago suggested that the quality of the editing subsystem (in terms of the user interface, speed, and capabilities) is as important to the overall flow of information through a GIS as the choice of a digitizing system.

6.4 Merging

With some modern geographic information systems, the essential elements to digitize from a source map or graphic are the points. If the system describes lines and polygons hierarchically, based on points, we must be able to build these hierarchically-defined spatial objects. **Merging** is the process of building more complex objects from the elemental points. Based on the data acquired during a digitizing session, vectors are constructed by connecting the appropriate points, and polygons constructed by linking the appropriate vectors. During this process, attribute values and non-spatial information can often be entered.

An important component of the merging process is the identification of objects that are very close to each other. For example, during digitizing we may have entered a line and a point that are less than 1 millimeter apart. By specifying a **tolerance** of 1 millimeter, we are instructing the system to "snap" the point to the line. Similarly, we may wish to have the system connect line segments that have end nodes within a specified tolerance. These processes have unfortunate side effects, however. Consider snapping the ends of two lines together: the result is that the location of the joining node is between the original line ends. If we now identify a third line, whose end is within the snapping tolerance of the same joining node, the system may again move the stored location of the joining node. Thus, the node locations may migrate through time!

Many of the consistency checks mentioned in section 6.3 can occur at this point in the operation of a geographic information system. This is often the procedure for systems based on arc-node types of vector data structures. This process can be considered developing a topological structure from the raw input data. The importance of an effective man-machine interface, where the computer hardware and software provide a productive working environment, cannot be overemphasized.

6.5 Edge Matching

Often, a region of interest is not completely contained on a single map sheet or a single aerial photograph. Similar problems occur with digital datasets that are stored with arbitrary boundaries (as in the quarter-scene format used in some satellite data, or when dataset coverages correspond to municipal or other political boundaries). In these cases, we must be able to extract information from each of the relevant map sheets, and process the information so that the edges between map sheets disappear to form a seamless dataset. This is particularly important when we are working with digitizing tablets to create digital databases from maps.

Consider the example sketched in Figure 6.10, where a region we wish to digitize into a GIS dataset lies across the boundaries of two map sheets. In this example, we see a water body and roads, both of which cross the artificial boundary between the two map sheets. One option is to place both map sheets on the digitizing tablet or scanning system at the same time, after manually creating a mosaic of the two. This is often impractical, both because the resulting composite sheet is too large for the available digitizing equipment and storage facilities, and because this can destroy the maps for other uses. The more common procedure is to digitize or scan each sheet separately.

Frequently, when the two map sheets are digitized or scanned separately, features that cross the boundary do not align properly (Figure 6.10b). This distortion can come from several causes. Even when maps are printed with no discernible error, the physical size of the map can change with temperature and humidity -- as mentioned before, this can be a significant problem with maps printed on paper. Errors at the margins can also be caused by georeferencing errors during the digitizing process, extrapolations and numerical round-off errors in the georeferencing algorithms, accuracy errors in the digitizing tablet itself, and slivers and gaps caused by overlapping map coverage. In the figure, it is not possible to simply move one of the maps to make both roadway and the water body match properly across the map boundary.

Under perfect circumstances, problems related to map shrinkage would be solved before entering the data into the GIS database. However, this is not always possible. There are two families of adjustments to correct these errors at the edges between map sheets or between different digital data files. In the first, the analyst manually adjusts the locations of points and vectors to

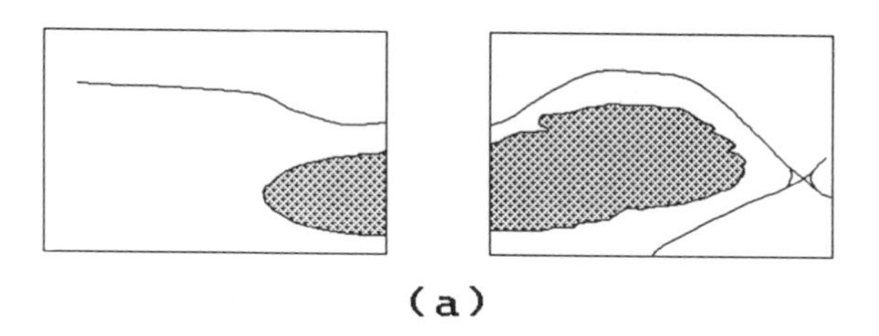

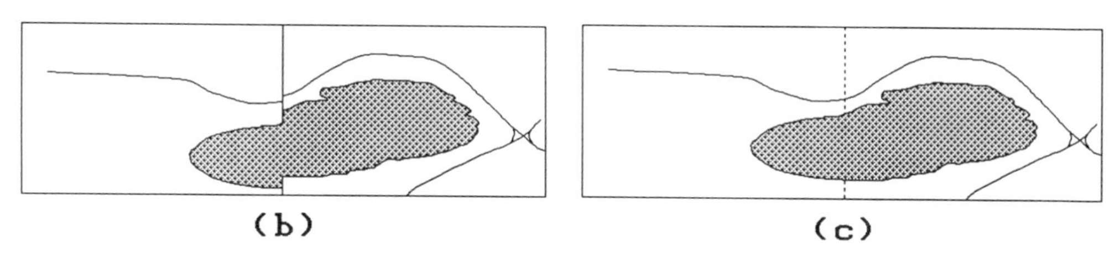

Figure 6.10 Edge matching. (a) two map sheets, (b) two sheets brought together showing discontinuities, (c) derived single sheet with edges adjusted.

maintain the continuity of the dataset. A graphics device is used to display the general area of the boundary between data sets, and the analyst uses "cartographic license" to manually adjust vectors that cross the boundary. This is the case in Figure 6.10c. Such a process implies that we have less information about the spatial data in question than we would prefer; be careful! In the second family of adjustments, automated means are derived to reduce the edge effects. Line attributes and the spatial distribution of the lines on either side of the boundary are first matched. Then, appropriate locations from the line on each side are modified slightly to match exactly at the boundary. The latter is much like the adjustment a surveyor applies to balance traverse data (for details, see Anderson and Mikhail, 1985).

6.6 Rectification and Registration

Rectification and registration are based on a family of mathematical tools that are used to modify the spatial arrangement of objects in a dataset into some other spatial arrangement. Their purpose is to modify these geometric relationships, without substantively changing the contents of the data itself. **Rectification** involves manipulating the raw dataset so that the spatial arrangement of objects in the data corresponds to a specific geocoding (or geodetic coordinate) system. As an example, a rough sketch of an undeveloped lot, made on-site by a landscape architect, typically does not show objects in their true spatial relationships. Once the sketch has been brought back into the office, the sketch can be redrafted (perhaps as a overlay on a reproduction of an existing map that shows legal boundaries and easements) so that it conforms to a useful coordinate system.

Registration is similar to rectification, but not based on an absolute georeferencing scheme. Both sketches from a field team and uncontrolled aerial photography have distorted views of the Earth's surface. The registration process involves changing one of the views of surface spatial relationships to agree with the other, without concern about any particular geodetic referencing system.

6.6.1 Map Projections and Coordinate Systems

Before considering the details of rectification and registration, let us briefly review the problems of extracting spatial information from the Earth's surface. The Earth's shape is roughly spherical, deviating from a perfect sphere due to gravity, centrifugal force, the heterogeneity of the Earth's composition, tectonic forces, and the influence of man's activities. A spherical globe can be a reasonably accurate representation of the Earth, since the oblateness of the general form of the Earth is less than one part in 300 (Robinson et al., 1978). However, a globe is inconvenient to manipulate and store, and expensive. Therefore, we must be able to work with maps: flat representations of the Earth's surface.

There are an infinite number of different map projections, since there are an infinite number of ways to project the features of the Earth's surface onto a plane. A variety of projections are in common use, since no single projection can be perfect for all users (U.S. Department of the Interior, 1984). Furthermore, there are a variety of commonly used systems for specifying locations on the Earth's surface.

The most common way to specify locations on the Earth is the latitude/longitude system, in which locations are specified in terms of the angular deviation north or south of the equator (to derive latitude) and the angular deviation around the circumference of the Earth at the equator (measured most often from Greenwich, England, to derive longitude). Unfortunately, it is not always convenient to convert these angular specifications into such familiar things as distances and areas.

Another family of common reference systems uses distances from a specified reference point as the basis of all locations. Such systems of reference are called **plane coordinate systems**. The Universal Transverse Mercator system is the best-known plane coordinate system, and is based on a transverse Mercator projection. In this system, the projection is secant to the Earth's surface, to balance scale variations. Further, the UTM system divides the Earth's surface into zones that are 6 degrees of longitude wide. Each zone is numbered, and the quadrilaterals 8 degrees of latitude high within a zone are lettered. This scheme of lettering and numbering provides a simple mechanism for locating areas on a coarse grid. Precise locations on the Earth are described in terms of north-south and east-west distances, measured in meters from the origin of the appropriate UTM zone.

Map projections are categorized in several ways (Monmonier and Schnell, 1988). A primary differentiation of classes of projections is based on the geometrical model of the projection (see Figure 6.11). In the simplest **azimuthal projections**, the map is constructed by placing a plane tangent to a point on the surface of the Earth. Features on the map are generated by systematically transferring the locations of these features from the sphere to the plane. We can imagine a transparent Earth, with an azimuthal plane tangent to the north pole, and a point light source at the south pole. The rays of light from the source may be traced in a straight line through points on the Earth's surface, to the tangent plane. Distortions are minimal in the immediate vicinity of the tangent point. These maps are popular for air navigation and radio transmission, with the destination or origin placed at the tangent point, since great circles passing through the tangent will be straight lines on this map. A more sophisticated version of the azimuthal projection is based on an intersecting or secant plane. In this case, the intersection of the plane and sphere describes a small circle, and the patterns of deformation are different than those of a tangent plane.

Conic projections are based on a cone placed over the Earth, oriented so that the intersection of cone and the Earth forms one or two small circles. Polyconic projections involve the use of a series of cones, each used to map a small fraction of the surface of the globe. For cones oriented conventionally, the axis of the cone is the same as the axis of the Earth's rotation. In this case, the small circles at the intersection of the cone and the Earth correspond to latitude lines. This family of projections is commonly used for displaying areas that extend a greater distance in the east-west direction than in the north-

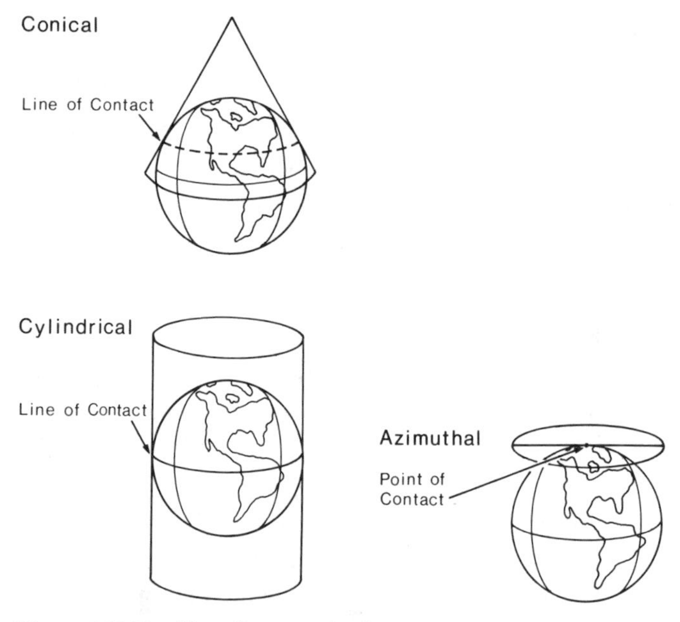

Figure 6.11 Families of map projections.

south, to take advantage of the orientation of the areas on the map of minimum deformation.

Cylindrical projections are based on a cylinder placed over the Earth. Again, in the simplest case, the cylinder is tangent to the sphere, with the intersection forming a great circle. In the most common orientation, the intersection is the equator. By reducing the diameter of the cone, the cylinder is secant to the Earth, and the cylinder intersects the Earth at parallel small circles.

The second set of classifications of map projections involves the patterns of deformations or distortions in the map. A projection is called **conformal** if there are no *angular* deformations. In order to maintain the accuracy of the angles, shapes of objects on the map are distorted. Thus, conformal projections are not good choices for the measurement of distance, since the scale changes across the map. Of course, the smaller the area covered by the map, the less important this problem becomes. The most well-known conformal map is Mercator's. A Mercator projection is based on the cylindrical projection, with mathematical adjustments applied so that all rhumb lines (lines of constant direction) appear as straight lines. Such a map has obvious applications in navigation.

Equivalent projections display relative *areas* correctly. Equivalence is an important consideration for data used as input to a geographic information system, so that inferences about areas are correct. By modifying the features on a map so that areas are correct, directions become unreliable. Common equivalent projections are the Albers conic, a frequent choice for maps that cover a large east-west distance, and the Lambert azimuthal equal-area projection.

Perspective projections represent the way a camera views its world. They are based on straight lines that pass from the object at a distance through a specified point of intersection at a finite distance from the plane of the projection (Thompson, 1966). Perspective projections present many problems to the user of a geographic information system. One of the simplest problems is that the scale in a perspective representation of the Earth's surface varies as a function of distance from the viewer: objects of the same size appear smaller when they are farther away. There are photogrammetric tools that can take this common but complicated view of the Earth and convert it to a planimetric representation (see, for example, Wolfe 1983). These are briefly introduced in section 6.8.

In practice, many geographic information system applications use plane coordinate systems such as the Universal Transverse Mercator system. These have the advantage that distances as recorded in the database correspond to distances one might measure on the (hypothetically flat) ground. When applications require the accurate measurement of area over very large distances, coordinates are often stored in the latitude-longitude system, with calculations based on an equal-area projection of the globe.

6.6.2 Exact Numerical Approaches to Rectification

There are instances when the input and desired output datasets are well-enough characterized that an exact numerical or algorithmic solution to the rectification problem exists. This is most often the case when the source data is a map of the Earth, in a common and specified projection and coordinate system, and the desired output is another map, in another coordinate system and projection. Another example is in the field of photogrammetry, where detailed calibrations are available for the camera system.

```
What is the INPUT Coordinate Type?
 0 = Geographic (Lon/Lat)
 1 = UTM                             11 = Lambert Azimuthal Equal Area
 2 = State Plane                     12 = Azimuthal Equidistant
 3 = Albers Conical Equal Area       13 = Gnomonic
 4 = Lambert Conformal Conic         14 = Orthographic
 5 = Mercator                        15 = General Vertical Near-Side Persp
 6 = Polar Stereographic             16 = Sinusoidal
 7 = Polyconic                       17 = Equirectangular
 8 = Equidistant Conic               18 = Miller Cylindrical
 9 = Transverse Mercator             19 = Van der Grinten
10 = Stereographic                   20 = Oblique Mercator (Hotine)
? 4

Lambert Conformal Conic

Enter LATITUDE of FIRST  STANDARD PARALLEL? 60
Enter LATITUDE of SECOND STANDARD PARALLEL? 40
Enter LONGITUDE of CENTRAL MERIDIAN? -105
Enter LATITUDE  of ORIGIN of PROJECTION? 0
Enter FALSE EASTING at CENTRAL MERIDIAN? 2000000
Enter FALSE NORTHING at ORIGIN? 0

Enter LATITUDE  of ORIGIN of PROJECTION? 0
Enter FALSE EASTING at CENTRAL MERIDIAN? 2000000
Enter FALSE NORTHING at ORIGIN? 0

SPHEROID selection:
 1 = Clarke 1866                     11 = Modified Everest
 2 = Clarke 1880                     12 = Modified Airy
 3 = Bessel                          13 = Walbeck
 4 = New International 1967          14 = Southeast Asia
 5 = International 1909              15 = Australian National
 6 = WGS 72                          16 = Krasovsky
 7 = Everest                         17 = Hough
 8 = WGS 66                          18 = Mercury 1960
 9 = GRS 1980                        19 = Modified Mercury 1968
10 = Airy                            20 = Sphere of Radius 6370977m

     >>> Enter spheroid number ? [1] : 1
```

Figure 6.12 Rectify menu entries. (Courtesy ERDAS, Inc. Copyright 1986. Used by permission.) This shows a session on an ERDAS system, running the function that converts vector data from one map projection to another. We present only the portion of the session in which the user specifies the map projection of the original dataset; specifying the desired projection for the resulting dataset is essentially the same.

When dealing with maps, we must have explicit information about the projection. For example, let's examine the information required to understand a common equal-area projection, the Albers conical projection. The first consideration, of course, is to find out which map projection is used for the map base, usually from information in the map legend or accompanying text. Next, determine the map scale, either from a printed notation on the map, or from a scale bar, or by using known distances between points to measure the scale. Finally, we must find the characteristics of this particular projection: in this case, the principal latitudes and the central meridian. The U.S. Department of the Interior reference (1984) provides very useful information on 20 common map projections.

A portion of some commercial software, based on this document, is shown in Figure 6.12. In this example, one can see the details required when specifying a map projection. From the menu, the user has selected Lambert's Conformal Conic projection. Latitudes of two standard parallels are required, indicating the small circles of minimum deformation. A meridian of longitude is selected, indicating the center of the map projection in the east-west direction. False northings and eastings are entered, at a specified geodetic coordinate location, to indicate the local reference axes. And finally, a choice must be made between alternative spheroidal representations of the Earth's shape. All of these components must be specified to determine the one specific map projection of interest, from the infinite number of possibilities.

An example of rectification, using an exact numerical approach, is shown in Figure 6.13. This example is taken from recent research in the boreal forests of North America. As part of a background study, we were interested in comparing several different maps that were developed to show the spatial distribution of species groups. As a part of this work, we needed to take the different maps and convert them all to a common projection and scale. The figure is a copy of a plotter output from a geographic information system, after a map depicting vegetation patterns in Canada was digitized (Rowe, 1959). The map is based on a conic projection; this is clear from the shape of the Canada-United States border, which corresponds over much of its length to the 49 degrees north latitude line. The research project required the rectification of this dataset to a latitude-longitude grid, so that it could be easily compared to several other datasets. The results of the rectification procedure, based on an algorithmic conversion between projections, are shown in Figure 6.13b.

The details of exact numerical rectification are based on equations of three-dimensional geometry, and are beyond the scope of this introductory discussion. For further details, see Snyder (1985). Modern geographic information systems usually include software to convert datasets among a set of specified map projections and geocoding systems.

Figure 6.13 Rectification. (a) conic projection, (b) data reprojected onto rectangular latitude-longitude grid.

6.6.3 Approximation Approaches to Rectification

Frequently, we do not have an exact solution to the problem of rectification, either because we do not know the details of the map projection in the datasets, or because the data do not conform to a standard projection or georeferencing system. An example of the latter situation may be found in an

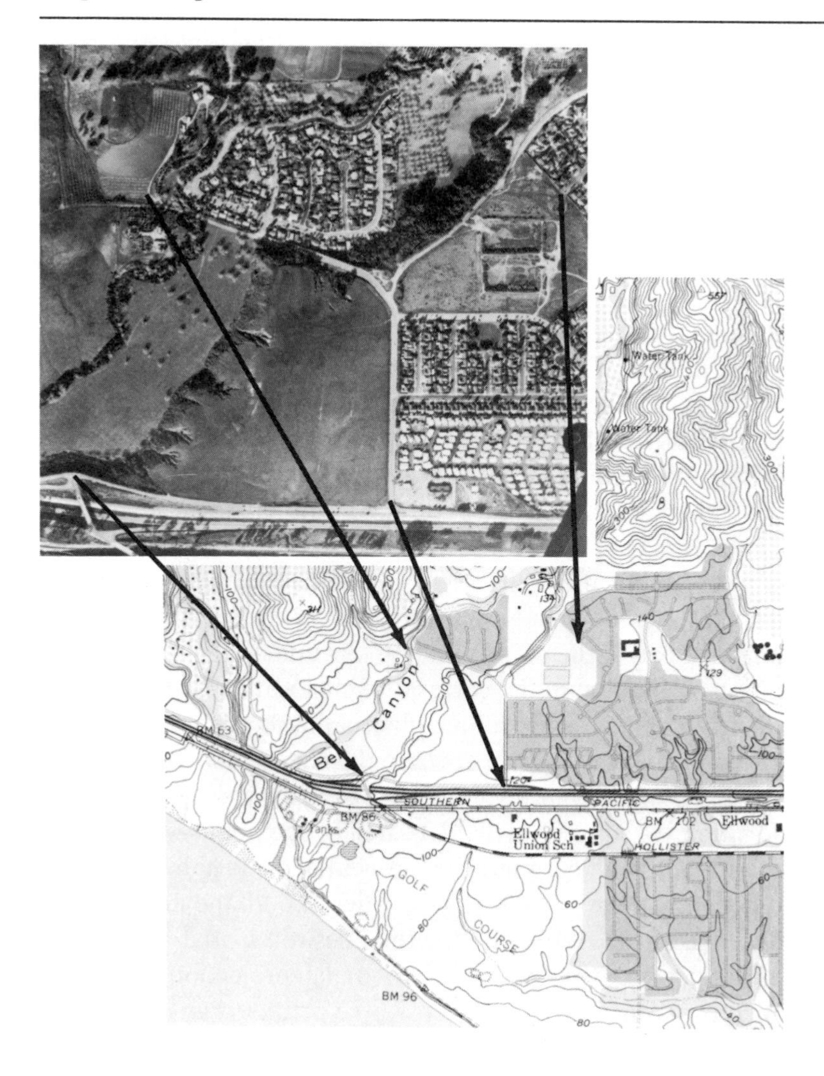

Figure 6.14 Representation of a rubber sheet operation.

oblique aerial photograph, in which the changes in scale across the image are due to the particular configuration of platform altitude, camera system alignment, and topography. A common approach, based on statistical operations, is called a **ground control point rectification.** In this technique, locations of objects in the source data (often based on the arbitrary reference grid of raster row and column) are compared to their locations on a trustworthy map (Figure 6.14). A statistical model is then used to convert the data into a new set, with the desired geometrical characteristics (Ford and Zanelli, 1985).

This procedure is often called a **rubber sheeting** operation, based on the following argument. Imagine starting with an uncontrolled off-nadir air photograph. This image will have many kinds of distortion, due to aircraft motion, perspective problems, the curvature of the Earth, scale changes due to topography, the distortions in the camera lens, and so forth. Imagine further that the film in the camera be stretched a great deal without tearing.

Next, place the film over a "correct" map of the Earth, selected based on the desired map projection and scale. Next, identify a location on the map that is easy to distinguish on the image, and run a pin vertically through the location on the image, down to the corresponding location on the map. This fixes the first location on the film to the corresponding location on the map. Next, identify another unambiguous location, place a pin through the photograph at this point, and run the pin down to the corresponding point on the map, stretching the film as required (Figure 6.14). This is repeated for each additional location as required. The identified image points which have been attached to the map are the **ground control points**, or **tie points**, and the number required depend on the geometric properties of both the desired map projection as well as the image. Note that the final desired spatial organization does not need to be a well-understood map projection -- we may well be registering one photograph to another, where neither is an orthographic representation of the Earth's surface.

The rectification process involves building a numerical coordinate transformation between the original image coordinates and the rectified (or "true") coordinates. Frequently, these transformations are based on polynomials, whose coefficients are computed by regression on the coordinates (see, for example, Welch et al., 1985). For example, we might use the pairs of coordinates (X,Y) as measured on the photograph versus (longitude,latitude) for a dozen ground control points to compute:

$$\text{longitude} = a + b*Y + c*X + d*Y^2 + e*Y*X + f*X^2$$

$$\text{latitude} = g + h*Y + i*X + j*Y^2 + k*Y*X + l*X^2$$

In these equations, a and g are called intercepts, and b to f and h to l are the coefficients, of the regression equations. The calculated intercepts and coefficients of the polynomials could be generated from the geographic information system itself or from a separate statistical package, and applied to the input (X,Y) data to produce (longitude,latitude) data. Moik (1980), among others, goes through the details of the process. Another numerical model for rectification uses only the nearest few ground control points to calculate the interpolation functions. This kind of procedure can provide reasonable performance, balanced by a need for many ground control points.

Once the appropriate numerical model for converting the source data to a useful geometrical form has been developed, the model is applied to the source data. When the data are in a vector form, the procedure is straightforward. The geodetic coordinates of the points or nodes are converted from the original reference system to the desired one, and all links between points and the accompanying vectors, polygons, and ancillary data are left in place.

The closest relative to a rectification procedure on vector data for raster arrays is termed **nearest-neighbor interpolation**. In this operation on raster data, we can convert the location of the center of a specified cell in the output image to a location in the input array, using the numerical model developed above. The attribute (or multispectral brightness) used for the output dataset is the precise value found in the cell nearest to the calculated input array location. In this way, the resulting data values in the derived dataset correspond exactly to values in the input data. This movement of raster element values from input to output arrays often causes a blocked appearance in the derived data (Figure 6.15), particularly when a rotation is involved.

A more complex means of rectifying raster data is to use a larger number of raster cells in the input dataset to derive the value of a single cell in the output data. In a **bilinear interpolation**, the values of the four cells in the input array whose locations most closely correspond to a cell in the output array are used. Linear weights are applied to the input cell values, based on their distances from the computed location that corresponds to the center of the output cell in the input coordinate system. This method requires substantially more effort than a nearest-neighbor algorithm, but the resulting dataset has a less blocked appearance. The linear interpolation performed in

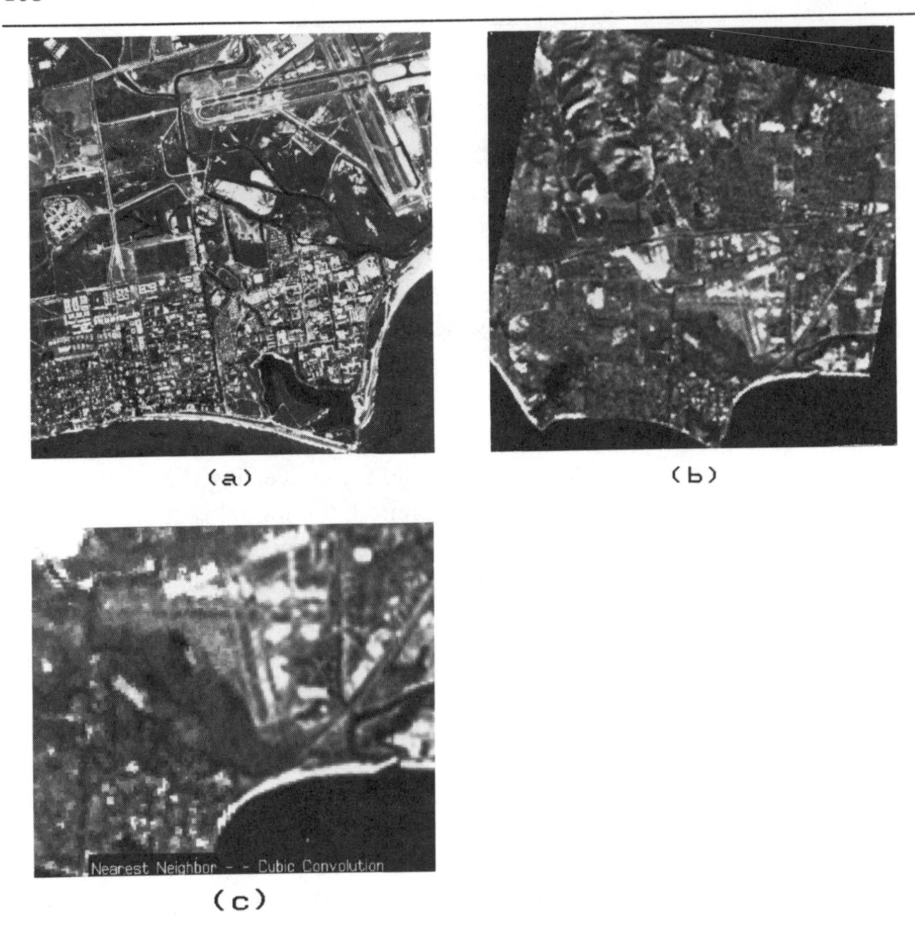

Figure 6.15 Interpolation in rectification operations. (a) original image data, (b) nearest-neighbor rectification, (c) nearest-neighbor versus cubic convolution.

this algorithm does cause a loss in dataset resolution, however.

Cubic interpolation is similar to bilinear interpolation, in that multiple data points in the original dataset are used to derive a single cell. However, in a cubic interpolation, a cubic polynomial is used to model the local behavior of the source data, and this local model provides an estimate of the characteristic value of the derived cell. This third algorithm often avoids the blocked appearance of a nearest-neighbor approach, and may degrade the spatial resolution of the data less than that of a bilinear interpolation. In exchange, there are increased costs of the computations due to the more complex calculations as well as the need to examine 16 cells in the input raster for each output cell.

For both bilinear and cubic interpolations, there is one important additional caution. In each case, the interpolation procedure averages a number of input data cells to calculate an attribute value for an output cell. Thus, the digital values in the derived data array may be values that do not appear in the input dataset. If the data values do not behave reasonably after a spatial averaging function, the derived datasets can be very misleading. Consider an area at the border of a lake, where we may use class 1 to indicate the sand on the beach, class 2 to indicate the paved roads and parking lots, and class 3 for the water. At one place along the edge of the lake, we find class 1, the sand, adjacent to class 3, the water. If we use an interpolation function to rectify these data to a specified map projection, which averages some of the class 1 pixels and some of the class 3 pixels, we may infer from the calculations that there is a class 2 pixel at the lake border -- in this case, we have spontaneously created a road that does not exist! In other words, for nominal and ordinal data types (as defined in Chapter 3), these numerical procedures are not appropriate since the averaging process is not well behaved.

A classic exercise in rectification and registration consists of taking a small dataset in one coordinate system, rectifying it to a second, and then rectifying back again to the original coordinate system. This can be very instructive to a new user, and can help the user develop an understanding of the capabilities and applications of the algorithms. For example, convert a small dataset in the UTM coordinate system to one in which distances are measured in miles and local magnetic north provides the reference direction. Such a conversion involves both rotation and scaling of the dataset. For vector data, the first (or forward) transformation should produce a reasonable representation of the initial data, and the second (or reverse) transformation back to the original coordinate system will be quite successful. For raster datasets, however, the choice of nearest-neighbor, bilinear, or cubic interpolation schemes will dramatically alter both the visual quality of the derived data as well as the success of restoring the data to the original georeferencing system.

6.6.4 Rotation, Translation, and Scaling of Coordinates

Rotation, translation, and scaling are mathematical components of the rectification and registration procedures discussed in the last sections. Rotation might be used to reorient data taken relative to magnetic north, so that the dataset is oriented relative to true north. Translation of the origin of a coordinate system is very useful, particularly when considering field measurements. A simple example of the use of translation of vector data involves a field survey, in which all recorded locations are determined relative to a local control point, such as the starting point of the survey. Thus, the survey coordinates are recorded as bearings and distances in meters from an arbitrary local origin. After the survey is completed, the locations can be biased by, for example, the Universal Transverse Mercator coordinates of the starting point, thereby changing all the measured locations from a convenient but arbitrary system of notation to a standard coordinate system. Scaling -- changing the values used to describe a location by a factor -- might be used to change from metric to English units of distance.

The computer graphics industry has developed a formalism for modifying coordinate locations to take rotation, translation, and scaling into account. While geographic data often requires higher-order adjustment as well, this matrix algebra development is useful. The following discussion is limited to two-dimensional data; it is taken mainly from Foley and Van Dam (1982), who also consider the three-dimensional case. Kemper and Wallrath (1987) also present a similar discussion in their article on geometric modeling in database systems.

Coordinate locations can be **translated** by simply adding the appropriate offsets to the coordinate pairs. Algebraically,

$$x^1 = x + \text{xoffset} \qquad y^1 = y + \text{yoffset}$$

This can be conveniently rewritten in matrix notation as:

$$\begin{aligned} P &= \text{original point location} = [x\ y] \\ P^1 &= \text{translated point location} = [x^1\ y^1] \\ O &= \text{offset vector} = [\text{xoffset}\ \ \text{yoffset}] \end{aligned}$$

and thus,

$$P^1 = P + O.$$

This operation is appropriate for either the nodes in a vector dataset, or the cell center coordinates in a raster dataset, subject to roundoff errors in the calculations.

Scaling involves applying scale factors to the coordinate pairs. The most simple use of a scaling function would be to convert between lengths in miles and lengths in kilometers, by multiplying the coordinate locations in miles by the factor:

$$\frac{(5280\ \text{ft})}{(1\ \text{mi})} \cdot \frac{(12\ \text{in})}{(1\ \text{ft})} \cdot \frac{(2.54\ \text{cm})}{(1\ \text{in})} \cdot \frac{(1\ \text{m})}{(100\ \text{cm})} \cdot \frac{(1\ \text{km})}{(1000\ \text{m})}$$

$$= 1.609\ \text{km/mi}.$$

Thus, scaling or multiplying the (easting, northing) coordinate pairs by the constant 1.609 converts displacements in miles to displacements in kilometers. In matrix notation,

```
S = scaling matrix = [ xscale    0
                         0     yscale ]
```

and thus,

$$P^1 = P \cdot S$$

In this example, the scale factor is the same for the x and y axes, which is sometimes called a uniform or **homogeneous scaling**. Under homogeneous scaling, shapes are preserved: circles remain round, and the length and width of squares remain equal. When the scale factors are not the same, we have **differential scaling**, and the proportions of objects are changed: circles become ovals and squares become rectangles.

There are a number of situations in which coordinate locations must be **rotated** through a specified angle. These include maps based on aerial photographs taken from a flight line that is not oriented due north. When rotating coordinates through an angle ***a*** about the origin (where positive angles are measured clockwise), the following equations are used:

$$x^1 = x*\cos(a) - y*\sin(a)$$
$$y^1 = x*\sin(a) + y*\cos(a)$$

Again moving to a matrix formulation:

```
R = rotation matrix = [ cos(a)   sin(a)
                       -sin(a)   cos(a)]
```

and,

$$P^1 = P \cdot R.$$

To tie these three geometric operations together, we introduce the concept of **homogeneous coordinates** (Foley and Van Dam, 1982). In homogeneous coordinates, it may be useful to apply a single scale factor to both the x and y values, so that a point is represented by:

$$P(x,y) => P(W^*x, W^*y, W)$$

The scaled x and y values can be converted to absolute cartesian coordinates by dividing the values by W, the scale factor, to recover x and y. Thus, points are now triples: *P*(X,Y,W). To show the value of this formalism, consider the scaling example above, where we converted our vectors from miles to kilometers. In the new notation system, all that is necessary is to change the value of W.

Since the elemental points are now three-element vectors, we must modify the transformation arrays above:

Translation:

```
                                 [   1        0      0
p1 = [x1 y1 1] = [x y 1] ·           0        1      0
                                  xoffset  yoffset  1 ]
```

Scaling:

```
                                 [ xscale    0       0
p1 = [x1 y1 1] = [x y 1] ·         0       yscale    0
                                   0         0       1 ]
```

Rotation:

```
                                 [ cos(a)   sin(a)  0
p1 = [x1 y1 1] = [x y 1] ·        -sin(a)   cos(a)  0
                                     0        0     1 ]
```

With this new matrix formulation, an appropriate sequence of translation, rotation, and scaling functions can be merged into a single matrix for the application at hand. For example, let's examine conversion from displacement in miles to kilometers, followed by a rotation of 8 degrees counterclockwise, to simulate a conversion from magnetic to geodetic north. The scaling operation becomes:

$$P^1 = P \cdot \begin{bmatrix} 1.609 & 0 & 0 \\ 0 & 1.609 & 0 \\ 0 & 0 & 1 \end{bmatrix}$$

As we have observed, in this case of homogeneous scaling (since xscale = yscale), we could have absorbed the scale factor into W. Considering the rotation operation, we get:

$$P^1 = P \cdot \begin{bmatrix} 0.9848 & 0.1736 & 0 \\ -0.1736 & 0.9848 & 0 \\ 0 & 0 & 1 \end{bmatrix}$$

Putting these two operations together in sequence, scaling before rotation, we can develop a single routine to produce the combined effect:

$$P^2 = (P \cdot S) \cdot R$$

$$P^2 = [\ x\ y\ 1\ ] \cdot \begin{bmatrix} 1.5845 & 0.2793 & 0 \\ -0.2793 & 1.5845 & 0 \\ 0 & 0 & 1 \end{bmatrix}$$

For vector datasets, it is a simple matter to use these operations on the points or nodes that are the fundamental units, and then construct vectors and polygons from the transformed points. For raster datasets, the operations may be substantially different. This is largely because with vector data, we are able to place objects at any arbitrary location in space, limited by our measurements. In raster datasets, a given raster cell's attribute value is often a spatial average of the characteristics of the primitive objects in the cell. In a raster array of population density, for example, a cell which contains a large apartment building surrounded by open space may have the same overall population density as a cell which is completely covered by a rectangular array of homes in a tract. For population data, in this case, which is a ratio variable, the averaging procedure is a simple mathematical function. For nominal and ordinal data, such as public versus private land ownership or classes of vegetation, the information categories and the procedures for aggregation will depend on the application. In an aerial photograph, if a sports complex is represented in a single cell of a derived brightness array, the brightness of the raster cell will be an average of that of the parking areas, the buildings, the stadium itself, the grass on the field, and so forth.

6.7 Interpolation

Many kinds of geographic data are often taken at irregular intervals in space. Elevation control points, demographic surveys, and vegetation biomass estimates all require a great deal of effort, and as such, it is usually impossible to make such measurements at all the desired places. There are valuable statistical principles that help in the design of such surveys, to maximize the amount of information gained for a given amount of effort; a brief discussion of sampling design is presented in Chapter 5.

However, there is often a need to estimate values for locations where there are no measurements. These estimates are generally based on the data that are available, plus a belief in (or preferably, an understanding of) the spatial variation of the phenomenon. The normal principle behind such estimation is that objects that are near are more important than those that are far away. For example, consider elevation data. If we have a small number of measured elevations in an area, we could construct a statistical model that lets us *interpolate* values of elevation at places where there are no measurements. Even better, if we know something about the local physiography and surface geological materials, we can do an improved job in our interpolations. We may be able to determine a value for slope above which the soil will slump. Thus, an upper limit to a physically reasonable slope can be determined. We may be able to statistically describe how rugged the terrain appears (Goodchild, 1980; Mark and Aronson, 1984). These kinds of local discipline information may permit us to narrow the range of possibilities and thus develop more reasonable (and thus, hopefully more valuable) interpolated estimates.

In any interpolation, the basic decision is to choose a model for the statistical relationships between data points. Two common models for such relationships are **weighting functions** and **trend surfaces** (Nagy and Wagle, 1979). Both functions are analytic descriptions of how to use known information to estimate that which is unknown. Weighting functions specify what each neighboring point's value will contribute to, that is, what **weight** it will have in, a determination of an unknown data point. A trend-surface calculation, on the other hand, is based on a two-dimensional polynomial approximation.

Consider a simple one-dimensional example, in Figure 6.16a. Imagine that we have two control locations, whose elevations are known, and we wish to estimate the elevation at a point on the straight line in between the two. A trivial model might be to calculate the average of the two elevations, and use this average to estimate the elevation of a new point between the two. This does not take into account the distances from the calculated point to the known points.

Another alternative, shown in the figure, explicitly takes into account the location of the new point. In Figure 6.16a, we have drawn a line between the elevations at the two known locations. This line describes a locus of points that correspond to an explicit interpolation model. This simple model is based on the following rule: changes in elevation between any two adjacent control points are linear with distance. Thus, it is a simple matter to calculate the predicted elevation at any point between the two control points. This second rule is more satisfying for interpolating elevation values than simply calculating the average, as we did at first.

Figure 6.16b sets up the complementary problem in two dimensions. We show three control locations (A through C), where we have measured elevation values, and a fourth (labeled D), where we wish to estimate elevation. Whether these data represent points in a vector database or the center locations of cells in a raster array, the general analysis is the same. First, we must identify the known points that are relevant to interpolating a new point. And second, we must decide on an analytic weighting function to calculate the new point. In this case, the three control points may have come from either of two simple rules: find the three nearest control points, or find all the points within a specified radius of the desired point. Common weighting functions include:

Arithmetic mean of the selected points.

This is appropriate when we believe that the spatial variation in the underlying data is small over the geographical neighborhood covered by the selected points.

Weighting the contribution of each control point by a normalized inverse of the distance from the control point to the interpolated point.

This is an extremely common approach, in which we assume that locations farther away are less important. The results of such a process for groundwater elevation data are illustrated in Figure 6.16d.

Weighting the contribution of each control point by the square of the inverse of the distance from the control point to the interpolated point.

This is a similar approach to weighting by the inverse of the distance, but in this case, locations further away are even less important.

A common geometrical model for interpolation involves building triangular networks, which are based on groups of three nearest control points. This model works as follows. To determine the estimated elevation at any point on the surface, find the nearest three control points. Determine the plane that goes through the three points, and then calculate the elevation of the desired point from the elevation on the plane that corresponds to the desired x-y position (see Figure 6.16c). This is directly related to the one-dimensional model we examined in Figure 6.16a.

Since these procedures involve search (to determine the appropriate points to consider for each desired location), distance calculations, and floating point arithmetic to perform the averaging functions, they can be computationally expensive.

The calculation of trend surfaces appears easy at first glance. An ordinary statistical program can be used to calibrate the model, where input data point values (such as elevation or population) are a polynomial function of x-y location values. A polynomial of an appropriate order is chosen to fit the points. Then, the function is use to estimate values at new points. In practice, it is important to understand that high-order polynomials tend to oscillate between fitted points, and this may produce unreasonable surfaces. One practical approach that avoids this kind of behavior is to fit low-order polynomials separately in each of a number of local neighborhoods throughout the data area. This kind of piece-wise approximation is often called a **finite element approach.** In this way, a separate polynomial is estimated for each local area, and known values distant from the area do not affect the local computations.

A family of interpolation techniques called **kriging** have been

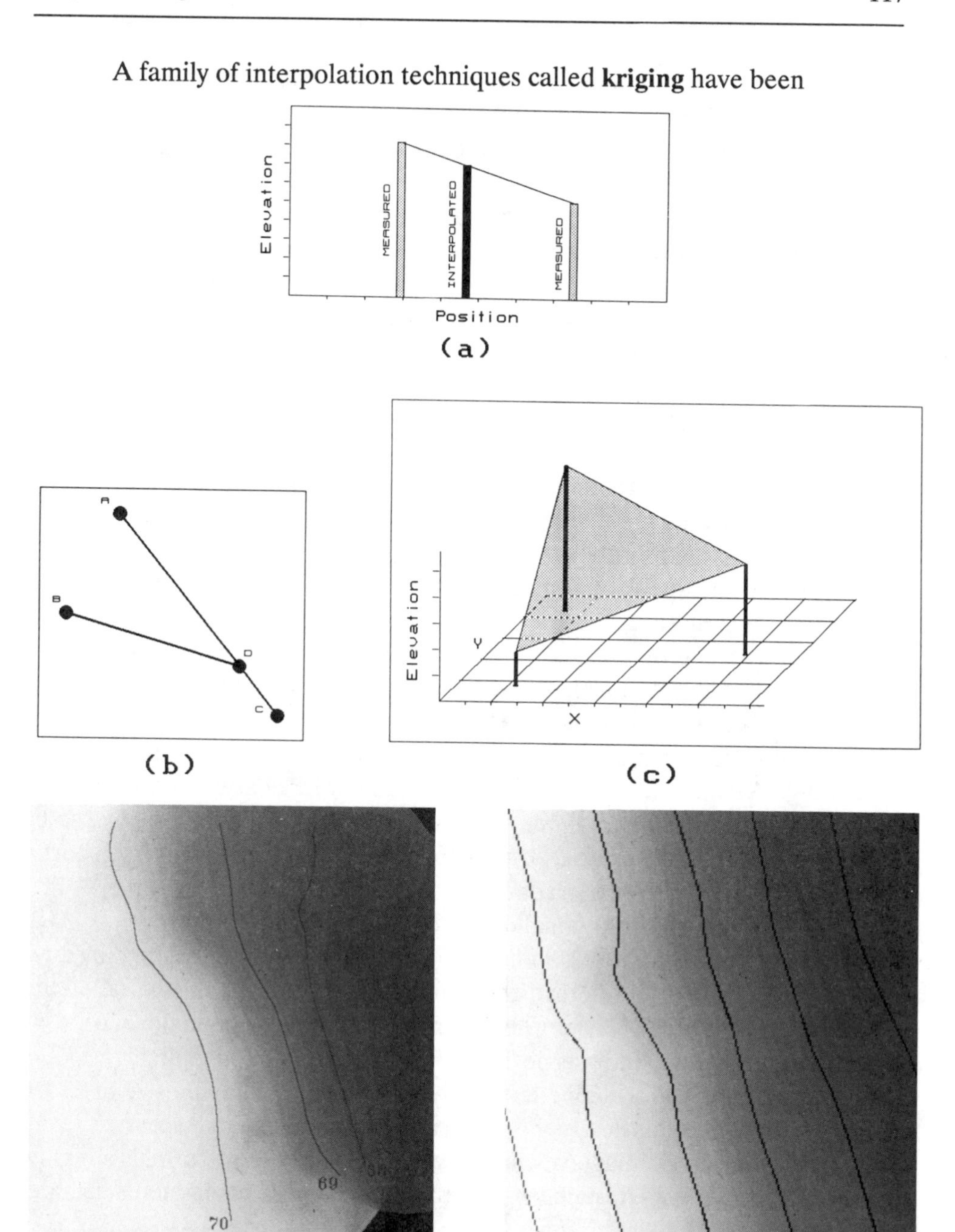

Figure 6.16 Interpolation. (a) one-dimensional, (b) two-dimensional, (c) three-dimensional, (d) distance weighted interpolation, (e) kriging of data in (d).

developed, which are designed to minimize the errors in the estimated values. The method is based on estimating the strength of the correlations between known data points, as a function of the distance between the points (Burgess and Webster, 1980; Webster and Burgess, 1980). This information is then used to select an optimal set of weights for the interpolation. Variations have been developed to include a trend-surface model component, which permits estimating values outside the area of known points (Figure 6.16e); this is not possible with a simple distance-weighted model as shown in Figure 6.16d.

6.8 Photointerpretation

An important but sometimes neglected component of data preprocessing in a geographic information system is **photointerpretation**: the process of extracting useful information from photographs and other images. Note that when we speak of photographs, we imply a data-acquisition activity that involves a camera or camera-like system, with lens and film. There are many other non-photographic systems that produce image data (see Chapter 10 for a number of examples) and we often apply the techniques of photointerpretation to these images as well.

Photography provides a tremendous resource for spatial data processing. Aerial photographs have been made of large areas of the Earth's surface, albeit principally in developed countries. Aircraft photography can be an inexpensive means to gather information about very large areas in a short amount of time. Also, photographs are efficient means of storing large amounts of data in relatively small volumes of space.

The development of new photographic systems has not stopped by any means. One of the most interesting recent developments is a device called the **Large Format Camera** (Doyle, 1985). This instrument has been flown on NASA's space shuttle, and was designed for high-precision mapping applications. It has a focal length of 30.5 centimeters, with a film size of 23 by 26 centimeters, and was designed specifically for mapping large areas.

The analysis or imagery, when viewed in its most basic form, is a highly complex process involving numerous components. These components include:

- a basic understanding of the specific application for which the imagery is being used,
- a general knowledge of the geographic area of the imagery,

- experience and expertise in image analysis techniques, and
- a sound understanding of image characteristics.

The essential tasks of photointerpretation are identification, measurement, and problem solving. Identification is recognizing the features of interest. Once these features are identified, it is possible to make measurements of such things as the distance between objects or the total number of features per unit area. Some of these measurements, of course, require that we understand the photograph's viewpoint, in terms of the variations in scale in the photograph due to the geometry of the camera/platform/Earth-surface system, the date and time of the image (to be able to calibrate shadows for sun angle, and to understand such things as seasonal effects), and so forth. The techniques of making accurate measurements from photographs are termed **photogrammetry**.

The process of photointerpretation is normally based on a systematic examination of the photograph, in conjunction with a wide range of ancillary data. The ancillary data are often very diverse, and may include maps, vegetation phenologies, and many kinds of information about human activities in the general area. The best photointerpreters have expertise not only in the delimited field of photointerpretation, but also in such related disciplines as physical geography, geology, and plant biology and ecology. In addition to these systematic components of image analysis, human interpretation also includes a significant perceptual or subjective component. Every analyst perceives features visible within imagery in a different manner. Also, the background of each interpreter is different in terms of knowledge, experience, and expertise in the various systems components and applications domains of image analysis. Consequently, human image analysis represents components of both an art form as well as a strict science.

The features found in a photograph are characterized by a list of image elements or primitives, which may be thought of as the basic characteristics of the image (Lillesand and Kiefer, 1987). These include:

Tone or Color:

The brightness (or color) at a single location on the image. The bright intensity of fresh snow differs from the darkness of deep fresh-water bodies (neglecting skylight reflected from the water's surface).

Size:

Size of an object can be considered in two different contexts. Relative size is a description that compares some objects to others in an image. Absolute size makes explicit use of our knowledge of the scale of an image; it requires detailed knowledge about the image that is not always available. However, in many images, we may be able to recognize objects whose size we know, and thus, we may infer the absolute sizes of other objects.

Shape:

The general form of an object is an important interpretation element. It is sometimes difficult to develop a taxonomy of shape, to be able to describe different shapes in a consistent way. The Pentagon Building in Washington, D.C., which appears as concentric five-sided shapes from the air, is the classic example of an object that may be identified by shape alone.

Texture:

Texture is based on spatial changes in tone. Texture in an image may be discernible at some scales, while at other scales, the spatial variations can not be resolved. Fine-grained texture implies large changes in brightness over small distances, such as the variations in brightness in a grassland. Coarse-grained texture implies relatively homogeneous regions with sharp discontinuities at the boundaries; areas of intensive agriculture often show a coarse texture from aircraft altitudes.

Shadow:

The shadows cast by objects can be important clues to interpretation. When data about a photograph include information for determining sun angle at the time of the image acquisition, shadows may provide excellent information about the height of objects. At the same time, shadows make it difficult to identify objects that are located in the shadows.

Pattern:

Pattern is discerned when there is predictability in spatial arrangements. The regular arrangements of trees in an orchard is an obvious example of pattern. Pattern is sometimes described as organized texture.

Site:

Site refers to geographic location. Site can be extremely valuable for understanding vegetation distributions in photographs, since vegetation species groups may only be found in certain kinds of sites (for example, swamps versus well-drained upland areas).

Association:

Association is based on the interrelationships of features at a location. For example, a fossil-fuel power plant is likely to be found associated with railroad facilities and power lines.

Tone can be considered the most basic of these elements, since without variations in tone, objects cannot be distinguished. Size, shape, and texture are more complex and are based on analyzing and interpreting individual features. Finally, shadow, pattern, site, and association may be considered the most complex, since they involve relationships between features.

As an example, we will briefly go through the image elements, based on the photograph in Plate 1. In this color infrared photography, the red tones indicate lush vegetation. Size and shape are diagnostics of several features, including the cars in the parking lot and the roofs of the houses. There is texture in the patterns in the sand on the beach, as well as in the vegetated median strip in between the lanes of the road. Shadows are evident adjacent to the houses; if you look closely, you can also see shadows from the street lights which (with information about the angle of the sun at the time of the photograph) can be used to estimate their elevation. The pattern of the lines in the image clearly tell us that there is a parking lot covering a large fraction of the image. The pattern of circular objects in the sand, combined with their size, suggests that these are fire rings (concrete rings in which it is permissible to light a fire). The association of the parking lot, fire rings, and beach suggests a recreational area.

The essential processes of human image analysis consist of the recognition of features and the formation of inferences (the latter through the synthesis of several elements of image interpretation). Features are recognized by the analyst's taking the elements of an image and applying to them procedures, techniques, and tools of analysis; it is through detection and measurement that objects and phenomena in an image are identified and their significance judged. Inferences are drawn typically through deductive reasoning using knowledge of both the specific application and the geographic area being examined.

A methodical approach is of particular importance to effective image analysis. By approaching an interpretive project in a systematic manner, the analyst is able to examine comprehensively all of the evidence present in the given imagery, minimizing the probability of an incorrect interpretation. Such an approach involves proceeding from the general to the specific, and from the known to the unknown. Implicit in this methodology is the development of parallel lines of reasoning through hypothesis testing, and verification of conclusions through convergence of evidence. The process is thus a deductive one, involving the recognition of basic features and a critical examination of evidence present within the imagery. A significant component of manual interpretation consists of identifying the most probable alternative from numerous hypotheses.

Photointerpreters commonly work with **interpretation keys**, which help to organize the process recognizing and characterizing features in images. One common form is called a dichotomous key, since at each step two alternatives are presented. A decision at a step in the key eliminates a number of possible interpretations, and the analyst then is guided to another decision. By stepping through these decisions, the analyst is led to a single (hopefully) correct interpretation of a feature. The interpretation keys of photointerpretation are closely related to those used in plant and animal biology to identify organisms to the species level.

In addition to evidence within the imagery itself, additional collateral or ancillary data are often highly useful, particularly when present in a geographic information system. Yet today this is seldom the case in manual image analysis; interpreters must search out or often create these data themselves. This can include ground truth data (such as field observations of species of vegetation), geological and geophysical information (including laboratory measurements of soil characteristics), and scientific literature (for example, a publication describing principal crops and forestry practices in a region).

Often these sources of ancillary data provide important clues that either substantiate or refute an interpretation. Also, the nature of the specific imagery used will have a large impact on the appearance of objects within an image. Factors such as sensor type (e.g., photographic, digital scanner, synthetic aperture radar), scale, spectral resolution, spatial resolution, sun elevation, time of year, and atmospheric conditions will have an influence on the utility of any given imagery for a specific application. Therefore, an image analyst must have a solid background in the characteristics of the type imagery being used, and in the potential types, sources and magnitudes of error that the imagery may contain or that can result from the variety of processing step which they may undergo. These are briefly discussed in Chapter 10.

Image analysis is thus an iterative process in which various types of information and the interrelationships between these and among the information types is examined. Evidence is collected, hypotheses are tested, and interpretations are made and iteratively refined in order to derive a correct interpretation. Often the derivation of new information requires the assimilation of lower-order image elements using convergence of evidence criteria. Consequently, the human analysis tends to be highly unstructured in nature due to the large degree of inter-dependencies between the various low-order image elements employed in the analysis procedure.

The output of this overall interpretation-analysis process is typically an overlay that contains the features or phenomena that have been extracted from the data. Less often the output of the analysis process may be produced in tabular or documentary form. Examples of these types of products might include a forest stand map, a product that can be used by resource managers to make decisions concerning harvesting scheduling. One aspect of such an analysis here might be timber-volume estimates for the stands identified on the image. As previously stated, however, the product of the analysis process is most often a map-like product because the geometry of the image format can produce distortions that must be taken into account when the analyst's results are prepared for input to a GIS.

In some instances interpretations from an image can be directly transferred to an overlay graphic of some kind. In other cases the product is often transferred to a map base using a direct entry process with local "eyeball" corrections for image distortions. The choice of a specific procedure to be used is governed by the overall accuracy, cost, and time requirements of the project being accomplished. Generally speaking, the higher the required accuracy, or the larger the area covered, or the more complex the analysis problem, the higher the cost. It is important to point out that an image-based analysis process is often a more cost-effective method of obtaining environmental information than relying exclusively on field work. Indeed, some types of information cannot realistically be obtained by any means other than aircraft or satellite data and photointerpretation.

There are a number of kinds of photogrammetric instruments that can be of great value during the preprocessing phase. Wolfe (1983) discusses their operation and the mathematics behind their design. We will briefly mention only a few of these devices.

There are a variety of simple devices for making measurements on photographs. Many of these are related to the vernier caliper, a common tool in a machine shop. This type of tool is used to measure length with very high precision. The complex part of the operation is to determine the relationship between the measurement of distance on the image and the true distance on the ground. The same comments are true for measurements on maps, since, as we have discussed, scale is not constant for many families of map projections.

Another family of photogrammetric instruments is frequently used to either develop or update planimetric maps. These systems project the image in the photograph onto a map or sketch, using components that can minimize distortions in the photograph. The zoom transfer scope is one such instrument. With this instrument, the operator views the map or sketch with the photograph optically superimposed. Direct revisions to the map can be made with relative ease.

A great deal of the information extracted from photographs is done through the stereoscopic viewing of sequences of photographs. By understanding the geometrical details of the camera system and the Earth's surface, it is possible to determine the horizontal and vertical positions of objects with very high accuracy and precision. The simplest devices for viewing pairs of photographs in stereo, called **stereoscopes**, effectively recreate the illusion of one's eyes being the same position as the camera lenses when the photographs were taken. In this way, the analyst can view the Earth in three dimensions. This ability to discern differences in elevation is a valuable technique in photointerpretation. Systems are available to aid in determining absolute differences in elevation, to quantify the differences in vertical position.

A family of instruments known as **stereoplotters** are used when pairs of photographs are the basis for developing accurate topographic maps. These complex systems are based on a pair of optical projectors, which recreate a three-dimensional view from a pair of photographs. A viewing system makes portions of the three-dimensional model visible to the operator, and a measuring system then records horizontal and vertical positions. Stereoplotters are the classical tools for compiling topographic maps. When we speak of an analytic plotter, we imply that portions of the stereoplotter system have been automated.

When an image correlator is a part of a stereoplotter system, the stereoscopic measurements once performed by the operator are also automated. The correlation process involves selecting portions of each image in the stereo pair, matching features in each, and, based on their horizontal positions, solving the parallax equations to determine both vertical and horizontal coordinates. These kinds of systems have had great success in those agencies with large-volume map production requirements. At this time, these systems cannot operate completely without human operator intervention.

After the analysis is complete and the product derived, the information must be input into the GIS. This is accomplished through the kinds of digitizing processes discussed in more detail in section 6.1.2.

Chapter 7

Data Management

As we discussed in Chapter 3, the **data management** element of a geographic information system is central to the overall system. The data acquisition and preprocessing phases we have already examined involve preparing the data for storage and use. The data management functions make the information available to the users.

There are a number of important components in any information system. The database itself is a structured collection of information. The tools of database management provide safe and efficient access to the database. The overall goal of data management is to provide users with such access without having to learn the details of the database itself. In effect, the database management system hides many of the details, and thus provides a higher-level set of tools for users.

An important distinction is made in the data management field between **logical data** and **physical data**. The way in which data appear to a user is called a *logical* view of the data. Through the database management software, the physical data themselves including the details of data organization as it actually appears in memory or on a storage medium, can be kept hidden from the user. As a brief example, consider a system that records publications of the faculty in a university. The faculty members may all be using the same computer system for their publications index, but each may keep track of her or his own publications in slightly different ways. One may use a fixed-length record organization, where information about each reference falls into pre-defined locations in sequence in a data record. The first 32 alphanumeric characters in the data record might be reserved for the first author's name, for example. Thus, the position of a piece of information has meaning.

Another may use variable-length fields, with tags to indicate record contents; the records in this case might have elements such as "YEAR=1988" and "KEY=HYDROLOGY." The elements "YEAR" and "KEY" are called tags, since they mark the fields so that we (and the applications software) know what the field contains. In some cases, there may not be rigid specifications for the lengths of these fields. Because of the presence of the data tags, the sequence of the data fields does not matter in this second case.

As long as all the required information for this application is available -- title of the paper, the list of authors, the name of the journal, etc. -- this can be managed. Thus, several different data files of bibliographic references, one for each faculty member, stored in possibly different physical organizations on different fixed disks, may be in simultaneous use. However, a modern database management system should be able to present a user with a single logical view of the dataset in a formatted presentation. The **database management system**, or DBMS, hides the physical details of storage and retrieval from the users.

Furthermore, different users may have different logical views of the stored data. In other words, the data does not have to appear the same way to all the users. Thus, logical views are dynamic in nature: they may change, depending on the user and on the application program. Such a system may also provide a mechanism for access control: anyone in the department may examine the contents of the bibliographic database we've described, but only the authors (or their designated representatives) may enter new references or modify old ones (for example, to correct errors, or to change an entry to indicate a paper formerly "in press" has finally appeared). In these ways, data management software allows a user to access data efficiently without being concerned with its actual physical storage implementation, and allows degrees of protection in terms of what a user may see, and what a user is permitted to do.

7.1 Basic Principles of Data Management

A database may be defined as data and information stored "more-or-less permanently" (Ullman, 1982), or as structured collection of information on a defined subject (Martin, 1986). A database management system is the software that permits one or more users to work efficiently with the data. The essential components of the system must provide the means to define the contents of a database, insert new data, delete old data, ask about the database contents, and modify the contents of the database. A DBMS is like a high-level computer language, in that primitive functions, such as the details of the operating system calls for opening and closing files, are hidden from the user.

Obviously, defining the contents of the database must be the first step in data management. The kinds of information required include:

Data format definition:

Is a specified data value to be stored as an integer -- and if so, how many digits are required for storage or presentation? Is the value a floating point number -- and if so, shall we use exponential notation? Is the data value a string of characters -- if so, is there a pre-defined format, such as those which are used for a date or telephone number?

Data contents definition:

The fields in the database are named. Useful names (such as CREATION.DATE or LATITUDE) are more helpful than arbitrary names (such as VARIABLE001 and X3).

Value restrictions:

Many systems permit the user to enter constraints on data values, which are then used to validate new entries. Variables such as date and time of course have constrained values: accidentally entering "month 13" or "63 minutes past noon" should be easy to prevent. Other values may have applications-dependent restrictions (such as the latitude and longitude values for a specified country), as well as default values.

These kinds of information about the data in a DBMS is normally stored in a **data dictionary**. The data dictionary is itself a database, and since it describes the contents of another database, we say that it contains **metadata**: data about data.

There are a number of important functions a database management system must provide:

Security:

As in the department-wide bibliographic database example above, all users should not have all modes of access to a database. Those without proper knowledge, or proper authority, should not have the ability to modify the contents of the database.

Integrity:

A DBMS checks elements as they are entered, to enforce the necessary structural constraints of the internal data. Data fields are checked for permissible values as described above; users (and applications programs) are forced to enter those data fields that are required; and so on. Furthermore, when the system has the responsibility of creating internal elements (such as pointers that explicitly provide links between data elements, or values that are derived from the entered data), they are created and checked for consistency. Checks for logical consistency are also required for the user's entries as well. Remembering the arc-node vector database structures discussed in Chapter 4, arcs are defined based on their end nodes. The database management system must prevent an arc from being described unless the related nodes exist. Further, editing functions must not be permitted to disrupt internal dependencies. Again considering an arc-node database, we must be prevented from deleting a node if there are still arcs that are defined based on that node.

Synchronization:

This refers to forms of protection against inconsistencies that can result from multiple simultaneous users. For a simple example, consider a database system for managing a collection of maps in a library. At the same moment, two users examine the database to determine whether a particular map is present in the collection. Both users are able to determine that the map exists, and they are each led to believe that it is currently in the library and available to be checked out. They each go to the library to check out the map, and one of the two users must be disappointed. A mechanism is required so that when one user is about to remove something from the collection, the other user is either warned, or prevented from accessing the information until the first has committed to the transaction. While most GISs rarely have more than one user accessing a portion of the database at a time, the problem is quite real in conventional database systems. Imagine one operator updating a land-use data layer, while another is trying to analyze the same layer: the analytic results will change through time, much to the chagrin of the second operator.

Physical data independence:

The underlying data storage and manipulation hardware should not matter to the user. In fact, under perfect circumstances, the hardware could be changed without users having any awareness of the change. This is a part of the distinction we made between logical and physical views of the database. It also permits system developers to write software without having to keep track of all of the details of physical data storage. Ultimately, this independence permits us to change hardware as needs and technologies change, without rewriting the associated data manipulation software.

Minimization of redundancy:

A desirable goal for a database is to avoid redundancy. When material is stored redundantly in a database, updating can be complex, since we must modify the same information stored in several places. If a data element is not changed identically in each of its locations, the database is corrupted. As we mentioned in Chapter 4, however, there are times when we optimize overall system performance by not following this tenet.

7.2 Efficiency

Efficient data storage, retrieval, deletion, and update are dependent on many parameters. Optimizing a database management system is a complex subject, discussed at length in many texts (see, for example, Chapter 6 in Martin, 1986). There are specialized systems in the commercial marketplace that are designed to handle certain kinds of database transaction problems, particularly in financial applications and travel reservation systems, with extremely high performance. We touch only on a few highlights of the subject. Two key elements to consider are the physical storage medium and the organization of the data (Calkins and Tomlinson, 1977).

The most common storage media in most computer systems are magnetic disks and tapes. As we write this manuscript, we are also beginning to see optical materials in commercial computer systems. Data on magnetic disks are typically organized as concentric circles of data bits, called **tracks,** with tracks divided into **sectors.** When more than one disk is held on a single drive spindle, all the tracks vertically above each other are termed a **cylinder.**

The magnetic heads that read and write the data on the disks are stepped across the disk surface to address the appropriate track, and then the system must pause while the relevant sector rotates underneath the disk head. Thus, the raw parameters of disk performance are governed by the speed at which the head assembly can move to a new track, and the speed at which the disk rotates. The speed of disk rotation, plus information about the linear density of data along the tracks, directly provides information about the ultimate data transfer speed to or from the disk. Since we can direct the head assembly to any track from any other, we call this a **random access** storage system.

If many users are simultaneously working with a single disk drive, they may cause a tremendous amount of head movement, as the system works to satisfy all the user's different requests for data. Two techniques are commonly used to improve disk system performance. A **cache** is a bank of high-speed memory used to store data that the system expects to need. For many systems, there may be little difference between reading a single sector of data compared to reading several sectors from a disk. Once the disk head assembly has moved to the correct track, reading sectors in addition to the one chosen may often take very little additional time. If the system can place one or more additional sectors from the disk in the cache, a subsequent request for data might be satisfied by the cache, without incurring the costs of head movement and disk rotation. This may be the case when sequentially reading from or writing to a file. If the data are organized on the disk such that caching is effective, system performance can be significantly improved, since data stored in the cache can be retrieved much faster than that stored on the disk.

A second way to improve system performance, particularly on multi-user computers, is to build intelligence into the disk controller to optimize a sequence of data requests. If there is a processor on the disk controller that examines all the pending data requests, it may be possible to schedule the requests in a sequence other than the order in which the request were made, to improve overall system performance. For example, the disk controller may be able to make use of knowledge of the distribution of data on the disk surface itself, in conjunction with knowledge of the current position of read/write head: to satisfy a sequence of data requests, read the data nearest the current head position before moving the head a great distance. This method may reduce the average time it takes to retrieve data over a large number of requests. Such a strategy must not be taken to an extreme; it must be ensured that all users have equitable access to the resources.

Magnetic tape storage is called **sequential access**, since the storage system must pass through the length of the tape to locate data elements, unlike the random access capabilities of magnetic disk systems. The industry standard tape records a linear density of 1600 or 6250 bits per inch, with nine parallel data tracks on half-inch-wide tape. Eight of these tracks are used for data storage, and the last provides some measure of error protection. Read and write speeds of 25 to 125 inches per second are common. Cache memory buffers are common in these systems, for a similar reason as in disk systems. Since the magnetic tape cannot accelerate and decelerate instantaneously, the cache can receive or provide data in very-high-speed bursts to improve system performance. Since these devices are inherently sequential, magnetic-tape drives are inappropriate when random data access is required. They are most commonly used for database backup, long-term storage, and data transfer between systems. We mention a caveat to long-term tape storage in Chapter 12.

According to Calkins and Tomlinson (1977),

> *The file structure should be determined by the complexity of the data structures, the data manipulations that must be performed, and the type of computer-aided techniques to be used, for example, regular computer job processing (termed batch) or interactive manipulation and analysis.*

A difficulty that arises here is based on the uses and users of the system. If a system is designed to support a single kind of processing problem, it is possible to define an optimum file structure, in terms of overall system performance. Design of such a structure would take into account the physical characteristics of the hardware, the design and operation of the software, and the data structure and its implementation. In contrast, when a wide variety of analytic operations are needed, supporting different kinds of users with different requirements and datasets, across several disciplines, such an optimization is much more difficult.

In the creation of a spatial database, it is necessary to provide modes of access for retrieval of both spatial and non-spatial (or attribute) information. Searches are conducted to locate features as well as sets of features. A GIS may be required to locate any of the spatial objects we mentioned in Chapter 1, or any of the components of the spatial database we discussed in Chapter 4. Some of the queries of the spatial database involve classes of features such as (from Calkins and Tomlinson, 1977; and Salmen, et al., 1977):

A single feature:

Find a second-level stream.

A set of defined features:

Find all second- and third-level streams.

An incompletely-defined feature or set of features (sometimes termed ***browse****):*

Find all features of (type = hydrography) in this region.

Features based on defined relationships within the data set:

Find all second-level streams above 3000 meters elevation.

A set of features in which the criteria are within another dataset.

For example, we may have an external database management system with both water-quality measurements as well as water-quality standards, in addition to a geographic information system with information about the stream network:

Find second-level streams with PCB pollutant levels above state water-quality guidelines.

Efficient data-retrieval operations (which of course are required during the processing phases in addition to an explicit query of the database) are largely dependent upon four items. The volume of data stored certainly affects data-retrieval speed, particularly when exhaustive search of the database is necessary. The method of data encoding can be very important to performance. This includes decisions about the types of variables to be stored as well as the way the values are stored. The design of the database structure can take advantage of knowledge of the types of problems to be solved, as we mentioned in Chapter 4. Finally, the complexity of the query directly affects the necessary calculations as well as the types and amounts of requests to be made of the database management system. These four concerns are all important to any evaluation of system performance.

In general, there are well-developed sets of procedures for reasonably efficient retrieval of non-spatial data. However, searches for spatial features or sets of features are considerably more complex, and optimizing systems' search performance under these kinds of conditions is an active field of GIS research.

7.3 Conventional Database Management Technology

At the present time, there are two principal approaches or models in database management systems. The following briefly discusses each of these in turn, to help users of geographic information systems to understand better a system's internal structure and capabilities. Modern textbooks in database management (such as Martin, 1986) treat this topic in much greater depth.

7.3.1 Relational DBMS

Relational database management systems have become extremely popular in recent years. The underlying model on which they are based is very easy to understand, and they are particularly well-suited for *ad hoc* user queries. As an example, consider the following data about the hypothetical golf course we've mentioned in previous chapters:

Hole	Nominal Length	Par	Area of Green	Total Area	Date Repaired
1	410	4	210	17690	10/87
2	365	4	160	20350	2/87
3	390	4	150	16980	2/87
4	150	3	75	4210	2/87
5	490	5	185	21760	2/87
6	340	4	95	11610	4/87
7	165	3	85	4500	10/87
8	475	5	120	18200	4/87
9	420	4	110	14540	10/87

In this example, the entire table is called a **relation**. Each row in the relation, which corresponds to the complete set of data about a single hole, is called either a **record** or a **tuple**. Each column within a record, which corresponds to a different kind of information about a given hole, is called a **field** or an **attribute**. For example, the record which describes the first hole stores five different facts about the first hole. These facts are the fields within the records, and the separate fields tell us that the first hole is 410 yards long, par for the hole is 4 strokes, the green is 210 square yards, the total area of the hole is 17690 square yards, and the hole was last refurbished in October of 1987. For those readers who are not golfers, the par for a hole is the estimated number of strokes required to play a specified hole, if one is skilled and makes no errors.

Asking questions of such a database in most cases requires that either a database programmer construct a custom software module for the application, or that the user learn a **query language**. The query language is the means of controlling the database management system, much like an operating system has its own language for controlling the computer. A sample query about the contents of this database might look like:

```
OPEN TABLE "FRONT-NINE"
      PRINT (HOLE, LENGTH) WHERE (PAR > 4)
```

The first command instructs the system to look in the appropriate part of the database (in this example, we are retrieving a set of records about a particular group of holes). The second command poses a specific question to the database manager. The system then responds with a table, listing the requested information (holes and their length) based on the specified constraint -- that the par for that hole is greater than 4:

```
QUERY #1: TABLE = "FRONT-NINE"
      PRINT (HOLE, LENGTH) WHERE (PAR > 4)
```

HOLE	LENGTH
5	490
8	475

One of the key characteristics of a relational database management system is that a single query can find more than one set of tuples, as we have seen in this example. More sophisticated queries might involve more than one attribute at a time:

```
PRINT (HOLE, TOTAL.AREA) WHERE (DATE = 2/87)
     AND (TOTAL.AREA > 10000)
```

This query might be used to begin planing the purchase of materials needed to restore some of the golf course, based on the elapsed time since the last refurbishment and the area which may need restoration. The database management system would respond to this query with:

```
QUERY #2: TABLE = "FRONT-NINE"
PRINT (HOLE, TOTAL.AREA) WHERE (DATE = 2/87)
     AND (TOTAL.AREA > 10000)
```

HOLE	TOTAL.AREA
2	20350
3	16980
5	21760

In later queries, we could ask the system to compute values derived from several queries, to generate such information as the total area of the greens (perhaps to estimate the amount of seed and fertilizer required in an annual maintenance budget). Note that these tables of data correspond closely to the organization of spatial and non-spatial data in several of the vector data structures described in Chapter 4, in particular, the arc-node (section 4.2.3) and relational data structures (section 4.2.4).

Chang (1981) notes that there are a wide range of functional query languages. At one extreme they are termed **procedural**, since the user must specify not only what is wanted from the database, but also the details of how to get it. Since the user must work relatively hard in this case, by specifying the procedures necessary to answer the question, the system processing may be relatively simple. At the other extreme are **non-procedural** query languages, which specify only the user's objectives. The system then has more work to do, since the user has supplied none of the implementation details. A popular model for a non-procedural interface to a database is termed **query-by-example** (or QBE), and is found in several commercial products. In a QBE system, queries are described by using the database interface to (in effect) describe the appearance of the output report. The system then develops a method of producing such a report, which essentially corresponds to the problem of writing a program in the database query language. Systems based on the QBE model can be extremely easy to learn.

As we discussed in Chapter 4, this relational database model is used in a number of operational geographic information systems. Lorie and Meier (1984) discuss some of the important issues in the use of such systems. In contrast to some operational GISs, relational database management systems often have well-developed capabilities for security, networked multi-user access, and concurrency control.

7.3.2 Navigational DBMS

In contrast to the relational database model described above, which is based on rectangular arrays of data, there are **navigational** database models which may be thought of as directed graphs (Martin, 1986). Usually broken down into **network** and **hierarchical** versions, navigational databases are also based on records, as discussed in the relational model in section 7.3.1. We note here that the term *navigational*, as used in this section, is not used in the usual sense of spatial analysis, where the goal may be route evaluation through a transportation network, or an evaluation of fluid flow through a plumbing system. Instead, we use the term as Martin does, to describe the following structure for an automated database management system.

In a navigational database, the relationships between different tuples are often displayed as links in a diagram. As an example, consider one organization for a collection of maps. We have separate kinds of records for storing information about the individual maps, about countries, about map series, and about the way maps are stored:

MAPS: Map Name, Area of Coverage

MAPSERIES: Series Name, Scale, Thematic Coverage

COUNTRY: Country Name, Area of Coverage

STORAGE: Map Name, Cabinet, Drawer

As is clear in Figure 7.1, there are a number of links between these different databases. When we retrieve information from this database, we **traverse** the links between the different record types; this is why we use the word *navigational* for these databases models. To retrieve a map of a particular country, we first examine the COUNTRY database to determine the geographic area covered by the country. We can also examine the

```
COUNTRY
  UNITED STATES          LATITUDE 19°N - 72°N ...
  CANADA                 LATITUDE 42°N - ...
MAPSERIES
  7.5-MINUTE TOPOGRAPHIC MAP, USGS
  1 DEGREE
MAPS
  GOLETA QUADRANGLE   119°52'30''W 34°30'N      7.5-MINUTE 1:24000
  CALFAX QUADRANGLE   120°07'30''W 36°22'30''N  7.5-MINUTE 1:24000
STORAGE
  ROW 6   CABINET 9   DRAWER 1
  ROW 6   CABINET 9   DRAWER 2
  ROW 6   CABINET 9   DRAWER 3
```

Figure 7.1 Navigational database model.

MAPSERIES database to find a map series with the appropriate themes (for example, we may be looking specifically for geologic maps). Based on the MAPSERIES database, we can follow a link to a specific MAPS database, where information about the maps for the particular series are stored. We then use the information from the COUNTRY database to find a specific map in the selected series. Finally, based on the unique map name, we follow a link to the STORAGE database to find exactly where the desired map is stored. Unlike the relational data model, in navigational systems we only retrieve a single tuple from a given query. In order to continue traversing links to find additional examples, we must specifically request that the system get the next relevant tuple.

Notice that there is a hierarchy in a part of the system we have described. The MAPSERIES database points us to one of a number of MAPS databases; thus, we might consider the MAPSERIES database as a **parent** of the MAPS databases, which would be called the **children**. Note also that there are many-to-many relationships in this example. One such instance is the relationship between MAPS, MAPSERIES, and STORAGE based on the map name.

The relational database model has become more popular than the navigational model in recent years. While the navigational model may provide a faster response time for predefined queries, the relational model is extremely easy to explain to a user, and is well-suited to *ad hoc* queries. Further, the relational query languages may often be easier to learn than those for navigational database systems.

7.4 Spatial Database Management

The functions of data management permit the efficient use of a database and are the entry points to hardware and software facilities. In a very real sense, there are two fundamental questions to ask of **spatial** data, no matter whether the data consists of a paper map stored in a cabinet or a digital file system within a geographic information system running on a multi-million dollar computer:

What is found at a given location?

Examples of this question might be:

What is the elevation of a specific geographic location?
At a specified coordinate, what is the soil type?

Are there examples of specified objects within a specified area?

Examples of this question might be:

Are there any fire towers in a specified region?
Where are any sources of potable water in a specified region?

These two fundamental questions are ultimately at the core of any geographic analysis.

In addition to answering these kinds of questions about spatial data, a modern GIS should possess a number of qualities that are common to all database management systems. These include:

Efficiency:

The storage, retrieval, deletion, and update of large datasets is an expensive process overall. These are the essential management functions for any database, and must be carried out efficiently regardless of physical storage device (fixed or removable disks, magnetic tape, or the new optical media) or database location (whether stored on a local computer, or accessible across a network or multiple-computer cluster).

Capability of handling multiple users and databases:

This means we must be able to support simultaneous access to the database by multiple users when required, as well as logically view the database as arbitrary subsets of the entire physical database.

Lack of redundancy of data:

Redundancy in a database is generally not desirable. In section 4.2.1, we discussed whole polygon vector structure, and noted that one of the problems with this data structure is its redundancy. In a database, storing values that are dependent on other stored values without explicitly keeping track of the dependencies can lead to disruption of the database. At the same time, storing and manipulating the dependencies, in addition to the data itself, increases the difficulties of working with the data.

Data independence, security, and integrity:

These three areas are, again, not unique in any way to geographic databases, but are guiding principles for any database management system. *Data independence* implies that data and the application programs that operate on them are independent, so that either may be changed without affecting the other. This is a fundamental consideration, and reflects a common database management point of view. Whenever possible, we must design systems in which components may be changed in a modular fashion, with minimum disruption of other components. *Security* refers to the protection of the data against accidental or intentional disclosure to unauthorized persons and protection against unauthorized access, modification, or destruction of the database. *Integrity* in this context is the ability to protect data from systems problems through a variety of assurance measures (e.g., range checking, backup, and recovery; Martin, 1975).

In conventional data processing terms, such support functions are referred to as **database management**. These principles are not firm rules that must be obeyed. Rather, they are useful guidance. In automated data processing systems, there are always tradeoffs to consider, balancing theoretical preferences against practical system operations. As we saw in an example in the data structures chapter, if we have a geometrical description of a bounded region stored in the database, we can always calculate its area or perimeter whenever required. However, if these values are needed repeatedly, it may be more appropriate to store the values permanently. If we do so, we are creating redundancy in the database, which will make data update more difficult. This will also increase the size of the database, since we are storing additional attributes. These costs must be balanced against the reduced processing needed during system operation.

Many retrieval strategies have been devised to facilitate manipulation and retrieval of large spatial databases. One of the most costly operations is searching the database. Some of the means to minimize the costs of search include:

Paging:

In paging (which is sometimes called **tiling**), the spatial database is subdivided into (usually rectangular) regions, which may not have to be the same size. In this way, operations on the database do not need to work with the entire database at once, but with a smaller set of pages (or tiles). This is reminiscent of dividing a study area among a set of map sheets. This internal stratification can dramatically improve a system's speed when responding to queries, but it must not burden the users. As the database grows in size (for example, when new subdivisions are built in a city), functions should be available to subdivide the existing pages into smaller ones, to keep the data volume in each page small, which facilitates rapid access. Complications will arise in this kind of system at the margins between the pages, where features may overlap the boundaries.

Retrieval within a specified region of interest:

In this case, the user specifies a spatial window of interest for the current set of queries. If portions of the database can be eliminated from consideration, the system can be much faster (and thus, ultimately, work with more data). To avoid problems of redundancy, this should be possible without copying the data in the region of interest to a new file.

Centroid as a search key:

If the centroid of an object's location is stored in the database, the speed of operations such as search for the conjunction between objects can be improved. Clearly, the centroid, as a single parameter description of an object, cannot completely characterize the object for all applications. However, indexing classes of objects by the location of their centroids can provide a dramatic help when searching through large spatial databases. Similarly, if the applications require them, other high-order descriptions such as perimeter, size, and indices of shape can also be stored to improve system response.

Hierarchical storage:

As discussed in section 4.1, hierarchical decompositions of space can offer significant performance improvements in some applications. By effectively having descriptions of space at different mean resolutions, it is sometimes possible to prune away regions very rapidly and to focus quickly on promising areas. At the same time, for some kinds of data, the hierarchical storage structures may provide mechanisms for space-efficient storage for some kinds of geographic data.

Topological encoding:

As discussed in Chapter 4, many modern spatial databases explicitly encode information about the spatial relationships between features. The DIME file structure, for example, has specific components to indicate which blocks are bordered by street segments. Since these relationships are explicitly stored, certain kinds of operations may be executed rapidly, in comparison to a data structure where resource-intensive search is required to uncover implicit relationships like adjacency and containment.

Such techniques are often developed and implemented within a specific operational environment (i.e., batch versus interactive, sequential versus direct access, mainframe versus minicomputer). Furthermore, particular encoding and retrieval techniques that take advantage of the data management and retrieval characteristics of a specific operating system may have been developed and thus, are dependent on a particular machine architecture. System design and retrieval techniques must be carefully examined whenever or wherever the transportability of a GIS to a different operating environment or operating system is a consideration.

Chapter 8

Manipulation and Analysis

We will discuss the analysis of data from data planes in a GIS without regard to the underlying data structure, (e.g., whether the data is stored in a raster or vector format). With algorithms available for both raster and vector data structures for virtually any analytic technique, as well as conversions from one structure to the other, which we have already discussed (Cicone, 1977; Peuquet, 1977; and Marble and Peuquet, 1977), a user of a geographic information system should be more concerned with requirements of the analytic task at hand, and less so with the internals of the database. Capabilities should also exist for the direct manipulation and analysis of the non-spatial attribute data files -- these are briefly discussed in the data management chapter.

The distinction between manipulation and analysis is rather arbitrary. The *Oxford English Dictionary* defines **manipulation** as "*the handling of objects for a particular purpose.*" In our case, the objects we are handling are the spatial datasets in the information system. The OED definitions of **analysis**, on the other hand, revolve around "*the discovery of general principles underlying . . . phenomena.*" Some of these procedures will modify the form or structure of a data layer in specific ways; some authors consider these **manipulations**. Other more general-purpose procedures can create data that may be non-spatial (such as the mean elevation in a region of interest or the number of linear miles of roadway in a city) or can create new information in a form similar to the original data (for example, creating a data layer of elevation in meters referenced to a local benchmark, which is derived from elevation in feet referenced to a regional benchmark). The development of new derived data layers, which may form the input to further analysis, is an important function of any GIS. Figure 8.1 outlines the procedures we will examine in the rest of this chapter.

- Reclassification and Aggregation
- Geometric Operations
    - Rotation, Translation, and Scaling
    - Rectification and Registration
- Centroid Determination
- Data Structure Conversion
- Spatial Operations
    - Connectivity and Neighborhood operations
- Measurement
    - Distance and Direction
- Statistical Analysis
    - Descriptive Statistics
    - Regression, Correlation, and Cross-tabulation
- Modeling

Figure 8.1 Data manipulation and analysis operations.

Many of these functions (Figure 8.1) are needed in the preprocessing phase discussed in Chapter 6, and we will try to be consistent about providing cross-references to the appropriate sections. Goodchild (1987) organizes spatial analysis functions along different lines. He discusses a set of basic classes of analytic functions, based on the inputs and outputs of the analysis process. For example, one of his classes is the set of functions that require only attribute information from a class of spatial objects. An example of such an operation would be to count the number of water wells in a database. Another of his classes involves functions that examine the characteristics of object pairs. An example of this class is a function that calculates distances between pairs of objects. This overall approach, developing a taxonomy of analysis functions, is reminiscent of the toolbox viewpoint discussed at the end of Chapter 3.

8.1 Reclassification and Aggregation

Frequently, the original data that is available to an analyst is not quite right for the task at hand. Two of the common problems are:

- The categories of information in the datasets need to be modified in some way to make them appropriate for the intended use. This arises frequently with both nominal and ordinal variables. Thus, the attributes themselves require modification.

- The data are not at the right resolution -- the spatial database is effectively at the wrong scale. This latter situation may be a problem for both raster and vector data structures, where the original specifications for data collection and database development did not capture all the spatial features required in later processing steps.

We will examine these two problems separately.

8.1.1 Attribute Operations

The original coding of attribute data may not be appropriate to later analysis tasks. For example, the categories of surface rock types in a geologic map may be too detailed for a particular purpose. It is easy to imagine an engineer needing to know whether a site is suitable for the construction of a specific facility. Rather than providing the list of the original geologic classes and information about the regions they cover, it would be more efficient to present the engineer with **recoded** data, in which the classes are described simply in terms of their suitability for the construction project: suitable, or unsuitable. Following this operation, the boundaries between the new suitability classes can be readjusted. This is often necessary after a recoding operation, since (in this case) a single suitability class may come from more than one original geologic class. The recoded data may have redundant boundaries that must be removed, so that a map that directly addresses the user's needs can be drawn.

Another example of attribute recoding comes from some of our recent work in forests. In Figure 8.2a, we present a map of tree species groupings, which may be based on both visiting the sites and aircraft observations. One of research goals in this kind of study is to determine whether satellite and aircraft multispectral scanners are able to distinguish between coniferous and deciduous trees, based on their multispectral signatures. In Figure 8.2b, we have recoded the polygons that were labeled by tree species into one of the two classes - *deciduous* or *coniferous*. Finally, in Figure 8.2c we derive the final map, by removing the redundant internal boundaries. This is a more complex operation in a vector-based data structure than in a raster-based one, since

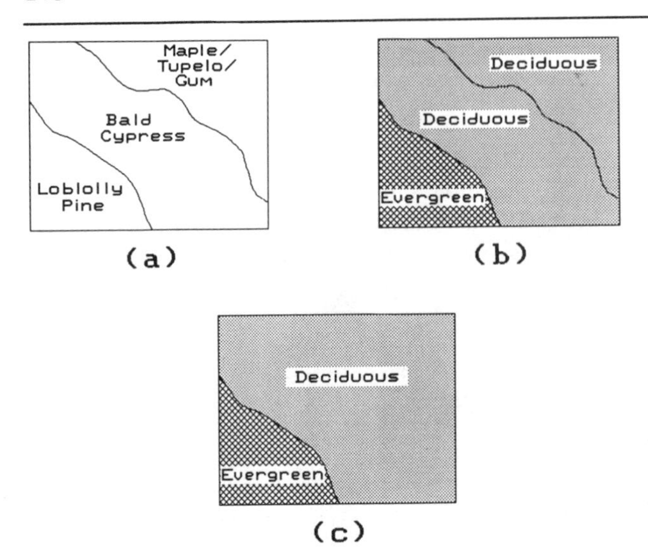

Figure 8.2 Attribute aggregation. (a) species map, (b) original map recoded to coniferous or deciduous classes, (c) redundant boundaries eliminated.

polygons are usually coded explicitly in the former and implicitly in the latter. In a vector data structure, eliminating a redundant line may require removing not only the line itself but also points which are no longer necessary, and then merging the appropriate polygons.

There is a much wider range of possibilities when working with more than one data layer. The **overlay** procedure is one of the simplest operations to understand; we've already seen an example in the golf course planning exercise in Chapter 1. When overlaying two data layers, a matrix can be a useful tool to describe the operation. As an example, we might consider what is called a *trafficability problem*: can a particular kind of vehicle travel across the terrain in an area, based on the slope and the type of soil? An example of this kind of analysis is found in Figure 8.3. The slope and soil data layers in this example are nominal variables, and form the inputs to the analysis.

In Figure 8.3b, we show the matrix of possibilities that is the basis of the traversal data layer. The columns in the matrix correspond to different soil types -- gravel, sand, and clay in this example. The rows correspond to different classes of slope -- level, moderate, and steep. The former is a nominal variable, and the latter ordinal. Each entry in the interior of the matrix tells us how easy it is for the vehicle in question to navigate a particular combination of slope and surface-soil type. For example, when the slope is moderate, the vehicle cannot travel over sandy soil, but can travel on a gravel surface. Figure 8.3c shows the resulting derived suitability map for traversal.

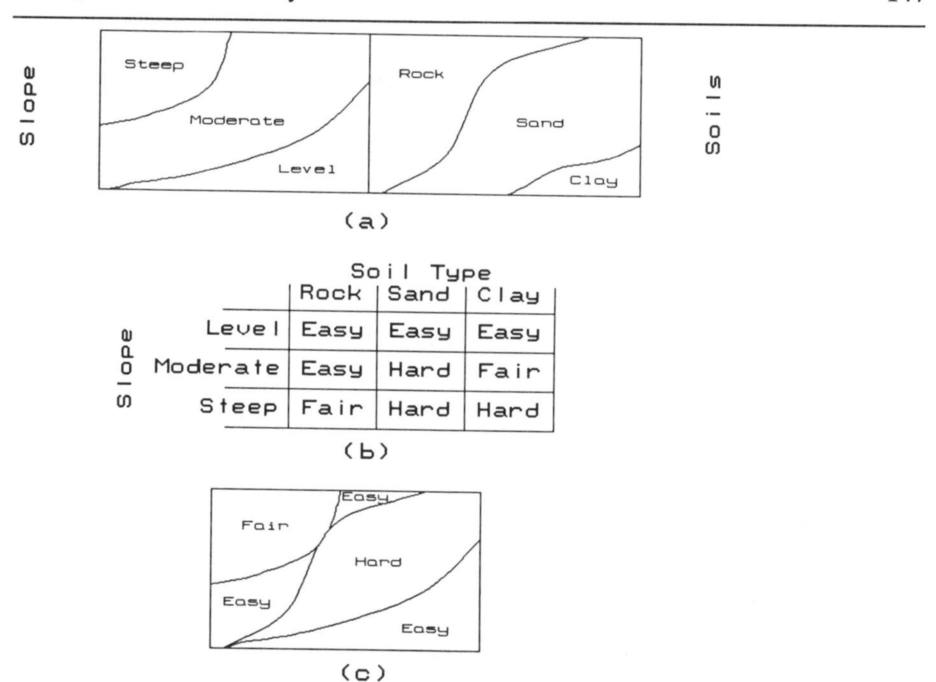

Slope	Rock	Sand	Clay
Level	Easy	Easy	Easy
Moderate	Easy	Hard	Fair
Steep	Fair	Hard	Hard

Figure 8.3 Overlay operation. (a) input map data, (b) conversion matrix, (c) derived traverse potential data layer.

In a raster-based system, each cell in the input data provides a soil-slope data tuple as input to the analysis matrix, which determines the class of trafficability in the output data. For example, in a band-sequential raster structure, this will be a simple sequential algorithm. We read the value of the first element in the soil array and the first element in the slope array, send these values to a routine that derives the resulting trafficability class, and send this derived value to the first element in the output array. This process continues through all the elements in the raster arrays.

In a vector-based system, the analysis is based on a polygon intersection algorithm, in which new polygons are created as needed, and redundant boundaries (as illustrated in Figure 8.2) are eliminated. Once the attribute lists of the new polygons are created, we pass through the attribute lists (the polygon file in a simple arc-node structure, or the polygon attribute file in a relational structure). For each polygon, the soil and slope attributes are sent to a routine as before, to derive a new trafficability value for each polygon that may be stored in a new column in the attribute list. Note that the vector case requires a relatively complex geometrical operation to derive the intersected polygons, and with it the creation of appropriate new nodes and arcs and their combined attribute values. This geometrical process is not necessary in the raster case. On the other hand, we generally have few polygons whose attributes are used to derive the desired answer for each polygon, in comparison to the relatively large number of cells to be processed in a typical raster database.

It is easy to see that this process can be generalized to larger numbers of data layers -- where in addition to slope and soils, a decision to drive a vehicle through an area may also depend on the vegetation cover (dense tree stands are hard to traverse, while open grassland is easy) and soil moisture (wet sand is hard, dry sand is easy). Furthermore, there are strategies that include the contributions of continuous variables (both ratio and interval), such as cost and time. Often, the manipulation of multiple data layers in this fashion is done in a stepwise manner: two input data layers combine to form an intermediate layer, and this intermediate layer is combined with the third source data layer to form another intermediate layer, and so forth.

Another form of the overlay process is found when the values of the original data layers combine mathematically to produce the output decision data layer. This is common when the data are ratio variables. Consider two data layers that may have come from an analysis of a natural forest: layer 1 contains polygons coded for the number of plant species found in this polygon in the understory, and layer 2 contains polygons coded for the number of species that make up the canopy. These two data layers may have come from separate field studies. To arrive at the total number of plant species per spatial unit, we in some sense "add" the two data layers. With raster datasets, one can easily imagine adding the values in each of the data layers, cell by cell. The software is more complex when we are dealing with a vector data structure, but the principle is exactly the same. Interior boundaries are created in the derived data layer wherever necessary, as described in the last example, and the attributes from the two input layers are added.

Logical (or Boolean) operations have many uses as well. Consider an example in which there are restrictions on construction and new development, based on specific characteristics at the site. New construction might not be permitted in any plot in a proposed subdivision where either (1) archaeological artifacts have been found, or (2) nesting sites for rare or endangered bird species have been discovered. When dealing with this kind of analysis, one data layer is normally coded for the presence of artifacts, and another layer coded for the presence of target endangered species. The decision for construction opportunities is based on the Boolean operator

"**OR**": if a lot is known to have artifacts, OR to have the designated target species, construction is forbidden. The OR operator creates the desired output data layer from the two input data layers. As before, with raster data structures, the operation proceeds on a cell-by-cell basis; with vector data structures, the geometry of the output data layer must first be derived before the attributes may be combined.

In this case, we have combined the exclusions together logically. If there are additional factors that could deny construction, we can sequentially take them into account by operating on the latest intermediate or derived layer, together with the next excluding layer. Similarly, we can include other characteristics that are required of a site with additional Boolean operators. To finish this example, we would overlay the subdivision boundaries on the derived exclusion data layer, and eliminate the excluded lots from consideration. The automated mapping-facilities management example in Chapter 12 shows precisely these kinds of requirements for both inclusion and exclusion.

There are several other Boolean operators. The "**AND**" operator is used to merge characteristics that are required at the same time. For example, new construction at a location is permitted when the local environmental study is completed AND plans for meeting any specific zoning restrictions have been developed. An exclusive-OR operator, abbreviated "**XOR**", is used to determine when one condition or the other is met, but not both.

There is another set of techniques for developing thematic descriptions from multiple attributes. Termed **classification**, these techniques start with several data layers and, based on a family of statistical procedures, derive a single multi-valued layer. Classification is designed to locate and describe relatively homogeneous regions, and is frequently used with remotely sensed data. Two different approaches are in common use: supervised and unsupervised.

In a **supervised classification**, the analyst develops multivariate descriptions of the different classes of interest. In a land-use application, the analyst could statistically describe the characteristics of each class. A certain type of recreational area might be described in terms of:

- public land ownership
- presence of on-site parking
- presence of rest rooms with wheelchair ramps
- area of the site is greater than 10 hectares

- the site contains a water body larger than 1 hectare

This description is sometimes called a training class for this land-use type, and we might have one or more specific examples of this type in the database. Other classes are described in a similar fashion. The GIS is then used to assign all the regions in the dataset to one of the described land use categories. There are a number of different decision rules that may be used to make assignments between the different classes, when there is no exact match between the available categories and the characteristics of a parcel or other delimited area (in a vector database) or pixel (in a raster database). In addition, various measures are available to examine the multivariate distance between the specified training classes and the parcel in question, and then assign the parcel to the nearest or most likely class (for example, see Moik, 1980).

The procedure is called a supervised classification, since the analyst *supervises* the choice of the classes. A user-friendly way to develop the list of classes and their characteristics is commonly found in systems that use multispectral image data. In these systems, the area on the ground which the analyst chooses as **training sites** for categories of interest are indicated with a light pen or joystick. Next, the system determines the characteristics of these selected sites from the stored database, and then develops statistical descriptions of the class. These derived statistical descriptions then form the basis for the subsequent statistical classification.

Unfortunately, this is not as straightforward a procedure as it sounds. The problems are based on the problem of variance capture. When there are variations in the characteristics of the members of a class, it may be difficult to accurately and unambiguously describe the class as a whole. In the land-use class example above, the area constraints may not be exact; there may be recreation areas that are very much like those described, but slightly too small. In a multispectral sensor dataset, there may be water bodies with different levels of phytoplankton biomass, thus presenting the sensor with different amounts of green reflectance. A common practice in trying to minimize the problem is to develop descriptions of several training sites within a class, so that the variability in the class is explicitly considered. Many of the implementations of supervised classification permit the analyst to calculate a measure of confidence in the labeling of each polygon. In a two class example, if the probabilities of assignment of a polygon into one of the two classes are 92% and 8%, respectively, one has confidence in the result. If the probabilities of assignment are 51% and 49% for another polygonal area, however, the region may be placed into the same class, but the analyst is less comfortable with the reliability of the result.

In contrast, **unsupervised** classification is much like the statistical clustering procedures used in ecological studies (Noyt-Meir and Whittaker, 1977). In unsupervised algorithms, the computer attempts to find different areas with similar attribute relationships. These areas are then labeled as belonging to the same class. It is then up to the analyst to determine a meaningful label for the classes, usually by working with site-specific information (see, for example, Jensen et al., 1987). Again considering the land use example, the software could scan through the data for commonly occurring patterns of land ownership, parking facilities, presence of water bodies and their sizes, and total area of the parcel. When statistically reliable patterns are discerned in this multivariate data, the system can group together those areas with similar characteristics.

From a mathematical point of view, the system is attempting to locate areas in the multivariate data space where observations cluster. Such routines can be a good tool for summarizing complex multi-layered datasets, as well as a good exploration tool, to help build hypotheses about spatial pattern and structure. Some background in multivariate statistics can be a great help in the interpretation, however; these statistical procedures can often permit naive users to fool themselves.

8.1.2 Spatial Aggregation

Spatial aggregation involves increasing the size of the elemental unit in the database. Since there is rarely an explicit elemental spatial unit in vector datasets (except for considerations of accuracy and precision, and the concept of generalization discussed earlier), this function commonly applies to raster datasets where there is a predetermined cell size. However, the concept is useful even in vector datasets. There may be instances in which regions of less than a specified area must be ignored for a particular application. In addition, a common operation with vector datasets is to merge adjoining polygons based on their attributes. These processes of changing the mean resolution of the data change the effective size of the minimum mapping unit.

The processes of spatial aggregation require the user to develop a **decision rule** for merging the attributes of the data. When the data describes

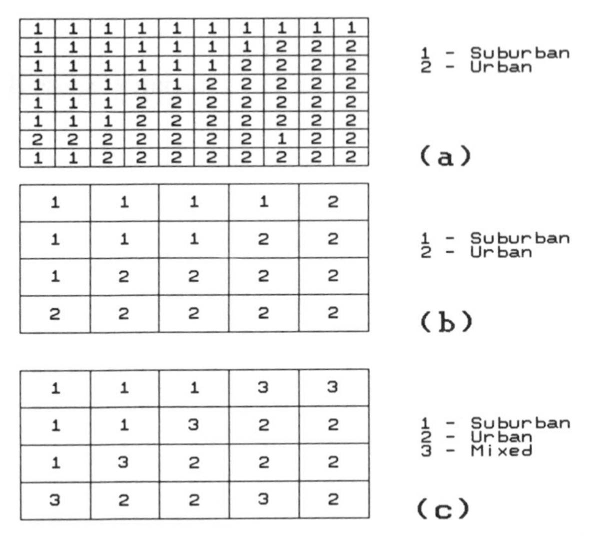

Figure 8.4 Spatial aggregation of categorical data. (a) original data layer, (b) aggregation based on a majority rule, (c) aggregation with the creation of a new mixed class.

discrete categories of information, such as land-cover classes, majority or plurality rules are normally used to derive the new category for the aggregated data from the old (Figure 8.4b). The example in the figure shows an urban area bordered by a suburban area. In this case, we are increasing the linear dimensions of the raster cell by a factor of exactly two, so that four original cells cover the same area as a single derived cell. During the process, the category that covers the largest fraction of the resulting aggregated area determines the attribute label for this new area; when there is a tie, we have specified that *urban* takes priority. An alternative, which keeps more of the information about pattern in the data, is to instruct the system to create a new "mixed" class for the aggregated data when there is no majority (Figure 8.4c).

For continuous data, such as elevation or average rainfall, an averaging process can be used to derive the aggregated data from the original values in the appropriate pixels (Figure 8.5a). This spatial averaging is a kind of filtering process, in which only general trends (or **longer wavelengths**, from a spatial analysis perspective) remain in the data. As in any other signal filtering process, we could have specified a different averaging function than the straightforward arithmetic average we used in this example.

When the aggregation process for raster data does not involve combining an integer number of cells to form a single larger cell, the processes of spatial interpolation discussed in section 6.6.4 are used. These tools use different models of spatial change to convert between different spatial scales.

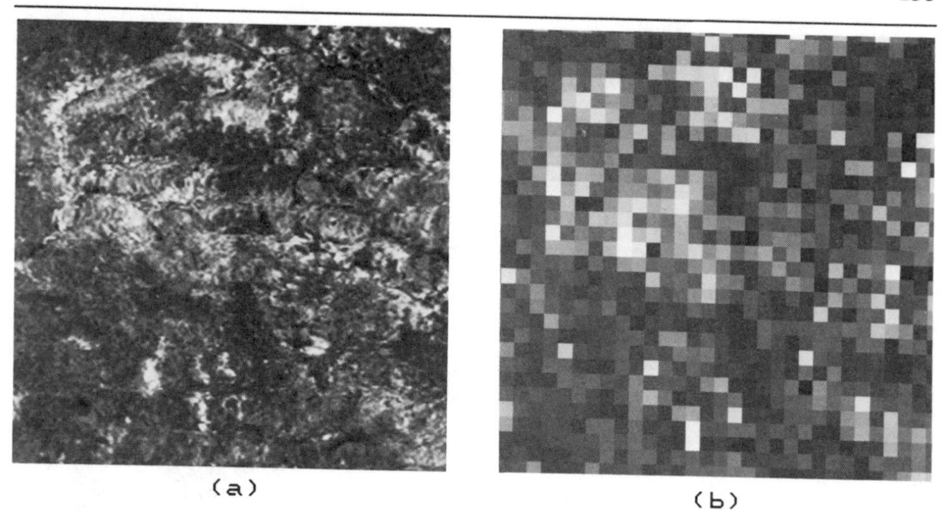

Figure 8.5 Spatial aggregation of continuous data.

This aggregation process is used frequently. There are often times when different data sets, each with its own spatial resolution, are all aggregated to a common scale before analysis. A simple example might be the integration of digital elevation data, with a raster cell size of 25 meters, with land-use data that has been processed to a 100-meter cell size. From a data management viewpoint, we must be careful to keep the original disaggregated data, even if it is not immediately useful, since it cannot be reconstructed from the aggregated data. The aggregation process is not reversible; that is, we cannot generally find a numerical operation that can identically recreate the original data from the aggregated data layer.

The spatial resampling examples we have considered in this section have only dealt with cases in which the desired spatial resolution is an integer multiple of the resolution in the original data. In Chapter 6, we discussed the more general case of **resampling**, to which this restriction does not apply. We must point out one important caution. When working with multiple datasets of different underlying spatial resolution (as we defined the term in Chapter 1), be extremely careful about the final results. From a simplistic but practical point of view, the end results can have no more spatial precision than that of the input data layer with the largest mean resolution element.

8.2 Geometric Operations on Spatial Data

This family of operations is important in both the preprocessing and analysis phases. In Chapter 6 we provide a detailed discussion of some of the uses of these functions in the preprocessing phase. In the following sections, we focus on using these tools during data analysis.

Rectification is the process of converting the spatial locations of objects in a dataset to those of a specified map projection. Rectification has been discussed in Chapter 6. Generally, it requires an explicit understanding of the geometrical characteristics of the source data, as well as those of the desired map projection.

Recent work with Landsat Thematic Mapper digital imagery (which has a pixel size of 30 meters) presents a useful case study of ground control point rectification. Welch et al. (1985) discuss the geometrical characteristics of TM imagery. Their analysis considered the numerical procedures necessary to register Landsat TM to U.S. Geologic Survey 1:24000 topographic maps. In their investigations, root mean square errors of less than 1 pixel could be obtained from as few as four ground control points for an area as large as 30,000 square kilometers.

As we discussed in section 6.6, there will often be cases in which the procedures in the previous section can't be used, because one or more of the spatial data sets does not represent a known map projection. There are many ways this problem arises. There will be instances in which a map-like product does not conform to a known map projection. Perhaps we have simply lost the details of the projection through the passage of time, or the cartographic process was not properly controlled, as is often the case in a field sketch map. A map may be based on measurements that have not been corrected for the effects of changes in elevation. Raster imaging sensors frequently present similar kinds of problems. These may be the result of a non-vertical imaging axis, distortion of many kinds in the optics and mechanical systems, curvature of the Earth's surface, and so on.

The tools for registering data layers under these circumstances have been discussed in section 6.6. In Figure 8.6, we illustrate a common example of this process, where a field sketch map, which does not correspond to any particular map projection, is registered to a map.

This brief discussion of coordinate systems has focused on point locations, and thus, primarily on vector datasets. When working with raster data organizations, there is an additional complication, since the area covered

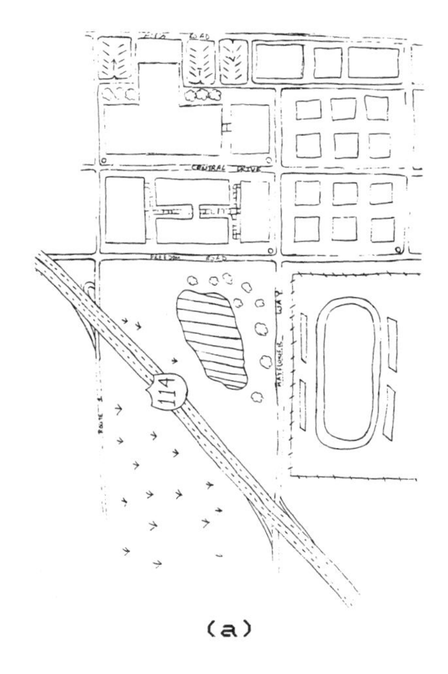

(a)

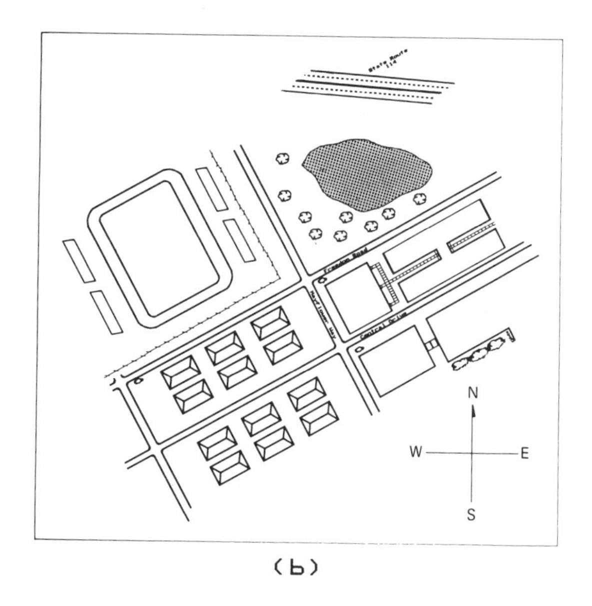

(b)

Figure 8.6 Registration example. (a) original field sketch map, (b) sketch map data registered to the base map.

by a cell in the registered data space will usually not coincide precisely with a cell in the original data space. This requires an interpolating function, as discussed in section 6.6.

The numerical tools of rectification and registration can be used in an inverse fashion, to intentionally distort a geographic dataset to make it communicate more effectively. For example, a planimetric dataset of landcover could be "distorted", to represent the perspective projection seen by an observer in a low-flying aircraft, or a motorist on a road viewing the watershed. Such a representation could be constructed using the tools of registration, and could effectively simulate a perspective view to a potential user. More sophisticated effects can be developed by integrating elevation data, and simulating view more accurately -- this is discussed in section 9.1.3. This specific example, simulating perspective views of a three-dimensional field, is an important component in the computer-aided design field. These kinds of operations, applying a specific geometric operation to analytic results, can also be used to mitigate problems in an output device, such as a projection video display with significant geometric distortions.

8.3 Centroid Determination

A centroid is a common way to describe the "average" location of a line or polygon. It can be defined physically as the center of mass of a two- or three-dimensional object. For a two-dimensional polygon as defined in a vector data structure, we could average the location of all the infinitesimal area elements within the polygon and could thus determine the coordinate location of the area's centroid. For an object defined within a raster, we could average the center coordinates of all the raster elements that make up an implicitly defined polygon, and could thus arrive at the object's centroid.

There is an easy way to think about a centroid. Consider taking a polygonal area on a rigid map, cutting it out along its boundaries, and then locating the point at which this two-dimensional object will balance on a single point. This point is the centroid of the area. Note, however, that there are cases in which the centroid position is not inside the object. For example, consider a U-shaped region; its centroid is between the uprights in the U, and thus is not in the area itself. Imagine a three-dimensional doughnut or torus: its centroid is a location inside the hole, not within the bounds of the solid object itself. However, most people would agree that even for these relatively difficult cases, the centroid location is a reasonable representative location for the objects. In some systems, after the centroid for a spatial object is calculated, the location is stored in the database as a searchable key. With the centroid as a location key, we can rapidly find objects that are relatively near a specified location. Thus, we can use the set of centroid locations as a spatial index for objects in the database.

8.4 Data Structure Conversion

Data structure conversion was examined in detail in section 6.1.1. It involves procedures for moving information in existing datasets from one specific data structure to another. It is frequently required in the preprocessing phase, to permit various datasets to be stored in a database that has a limited number of internal data structures. At the same time, there may be instances where, in order to run a particular analytic function, the data must be in a particular structure. This happens frequently when using external numerical models, as discussed in section 8.8. Thus, we must have access to the conversion functions even during analysis. In addition, we will have a need for these kinds of tools when generating output products as well, as discussed in Chapter 9.

8.5 Spatial Operations

Spatial analysis involves operations that consider the spatial arrangement of information in one or more data layers. They may be broken down into two families: operations concerned with **connections** between locations, and operations concerned with local or **neighborhood** characteristics.

8.5.1 Connectivity Operations

One of the classic functions of a geographic information system is to be able identify items in **proximity** to each other. This function has a wide variety of applications, including identifying impact and easement corridors, and in problems of site selection.

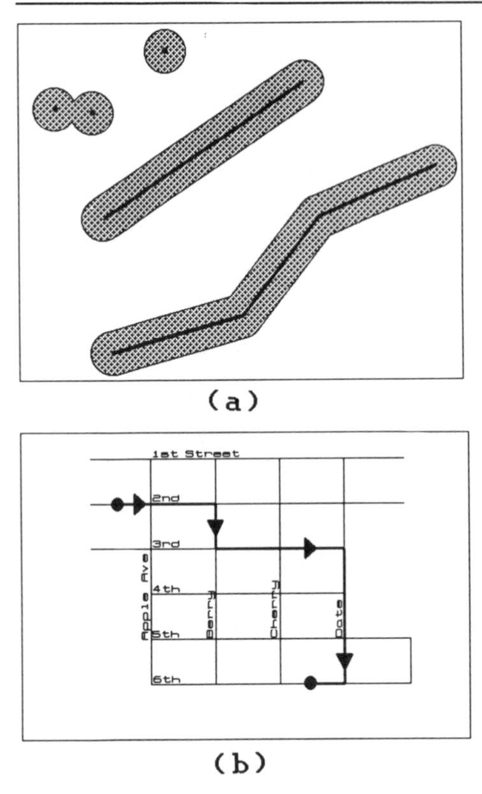

Figure 8.7 Spatial analysis. (a) proximity or buffer function, (b) network analysis.

In a site selection context, it may be important to locate near certain kinds of activities, and to avoid locating near others. If we can specify the appropriate distance criteria, the GIS can help narrow down the search for acceptable sites by taking a data layer that stores the locations of key activities, and deriving a new data layer that describes the impact zone around these activities. A hospital should be relatively near to major transportation routes to make it easy for people to get to the hospital, for example, but far from sources of noise and air pollution.

For a more complex example, consider the problem of residential development near an airport. Locations under the principal flight paths and those within a specified distance of the airport boundary may be subject to restrictions on the height of buildings and other facilities. If we have geographic data layers in which the flight corridor boundaries as well as the airport boundaries are stored, we should be able to calculate the locations of those areas that are subject to the height restrictions. Developing the boundaries of the region within the specified distance from the airport is called a **proximity** or buffer operation, and is a common tool in spatial analysis.

Proximity operations on vector datasets may require relatively complex geometrical operations. Locations within a specified distance of a point will form a circle; when the indicated points are closer than twice the specified distance, the circles will overlap (Figure 8.7a). Proximity boundaries on individual straight lines are relatively easy to develop and are more complex for either connected line segments or curves. Similar operations on raster data are easier to imagine, since we are effectively counting pixels away from the specified cells.

There are a variety of kinds of spatial analyses that are based on **networks**. Applications in this general area include optimum corridor or travel selection, and hydrology and flow functions. A complex but useful function found in some systems is to be able to identify the separate watersheds in an area, through run-off direction calculations that are based on a terrain description. Another interesting example of a network analysis problem is to evaluate alternative routes for emergency vehicles, based on a combination of total length of the route, plus anticipated congestion on surface streets, which is dependent on the time of day (Figure 8.7b). This is a complex problem in systems analysis, and is not usually the province of general-purpose geographic information systems. In some commercial systems, these kinds of applications separate software packages that interface to the GIS, for both access to the database as well as display functions.

8.5.2 Neighborhood Operations

There are a variety of different operations that are based on local or **neighborhood** characteristics of the data. In vector datasets, we often store information about nearby objects, such as polygons which are adjacent to each other. Thus, if we need to know which polygons are adjacent to a specified target polygon, line, or point, we can search through the data structure files in a straightforward fashion. For example, this kind of operation permits us to examine the various kinds of land-use surrounding a specific parcel which a client might wish to purchase.

In a raster dataset used to store many different kinds of variables, we can take the same approach. We can examine the attributes of the pixels surrounding a set of one or more target pixels, and look for such things as variability in certain attributes. On the other hand, when the raster contains values of an ordinal, interval, or ratio variable such as *per capita* income or rainfall, one of the simplest local operations to imagine, and an extremely common function, is computation of the local mean value in a single data layer. If we consider that every pixel represents the center point of a 3-by-3-pixel neighborhood, we can directly calculate the mean of this local area. The new set of mean values forms the derived data layer. This operation is a low-pass filter, in the language of signal processing. It will identify the general trends in the data, and reduce or smooth out the effects of single cells that vary from the local trends. This can be a very useful function when examining "noisy" data. Without considering the details, we note that a similar operation is possible in vector datasets. However, in this latter case we are examining the attributes of objects (for example, polygons) within a specified distance of a target object.

The **mean filter** we have just described is a special case of what is called **spatial convolution.** We have applied a set of weights to the raster cells in a local area, to derive a value representative of that area. In the case of the 3-by-3 mean filter above, we have summed the 9 values in the local neighborhood, and then divided by nine to preserve the local mean. Alternatively, we can weight each value by the constant 1/9, and then sum the weighted values. It is common practice to represent this operation by writing out the array of weights:

$$\begin{bmatrix} 1/9 & 1/9 & 1/9 \\ 1/9 & 1/9 & 1/9 \\ 1/9 & 1/9 & 1/9 \end{bmatrix}$$

This array is sometimes called the **kernel** of the convolution operation. If we were interested in the mean of a larger area, we could specify a larger spatial window -- perhaps a 5-by-5 cell region. By increasing the size of the window for this simple mean filter, we smooth the behavior of the data layer more dramatically. In other words, we attenuate the high-frequency information in the data more strongly. Figure 8.8a shows a raster dataset of infrared reflectance, derived from the Landsat Thematic Mapper satellite sensor (which is discussed in more detail in Chapter 10). Figure 8.8b shows the same data after operating the mean filter. By removing a portion of the "high-frequency energy" in the data, or in other words, focusing on the general (or "low-frequency") trends in this information, we get a very different picture of the spatial pattern in land cover and land use. More detail on various convolution operations is presented in Moik (1980); we will only discuss some of the most common techniques.

In direct contrast to the low-pass mean filter we have described, there are a family of edge-detecting or high-pass filters in common use. One frequent edge-detecting filter has the following kernel:

$$\begin{bmatrix} -1 & -1 & -1 \\ -1 & 9 & -1 \\ -1 & -1 & -1 \end{bmatrix}$$

As in the 3-by-3 mean filter above, the sum of the weights is 1, which suggests that the filter will approximately preserve the mean of the dataset. We point out that these operations are not well-defined for categorical data, and may produce results which are difficult to interpret. If these kinds of filters are applied to data that is not continuous and naturally ordered, such as classes of land use or lists of species, these operations may not make sense. Figure 8.8c shows the operation of this high-pass filter on the remote sensing data; note how it complements the low-pass filter shown in Figure 8.8b. This view of the original data emphasizes areas where there is change in adjacent land-cover types.

Another common spatial manipulation, which is directly related to the convolutions discussed above, is a **median filter**. With this filter, rather than calculating the mean value of a neighborhood, we calculate the median value. The median is the central value in a dataset, in that half of the data values fall above the median, and half fall below. This kind of calculation may be of particular value when the data has outliers or is otherwise not statistically well-behaved. This operation is another kind of low-pass filter, as it can remove the effect of outliers and can focus our attention on the general patterns in the data. This filter is popular with analysts who prefer non-parametric statistics (see Winkler and Hays, 1975), which are algorithms for analyzing data without regard to the underlying frequency distributions.

Texture transformations are another set of tools that are used to identify the spatial pattern in data. Texture filters are designed to enhance the heterogeneity of an area. A common algorithm for a texture filter is to calculate the standard deviation of the values in a 3-by-3 neighborhood. If the attribute values in this neighborhood are all similar, the standard deviation is small, and we say that this neighborhood has low texture or low variability. Where there are many different attributes in a neighborhood, we have high texture. These transformations are also used to find the boundaries between

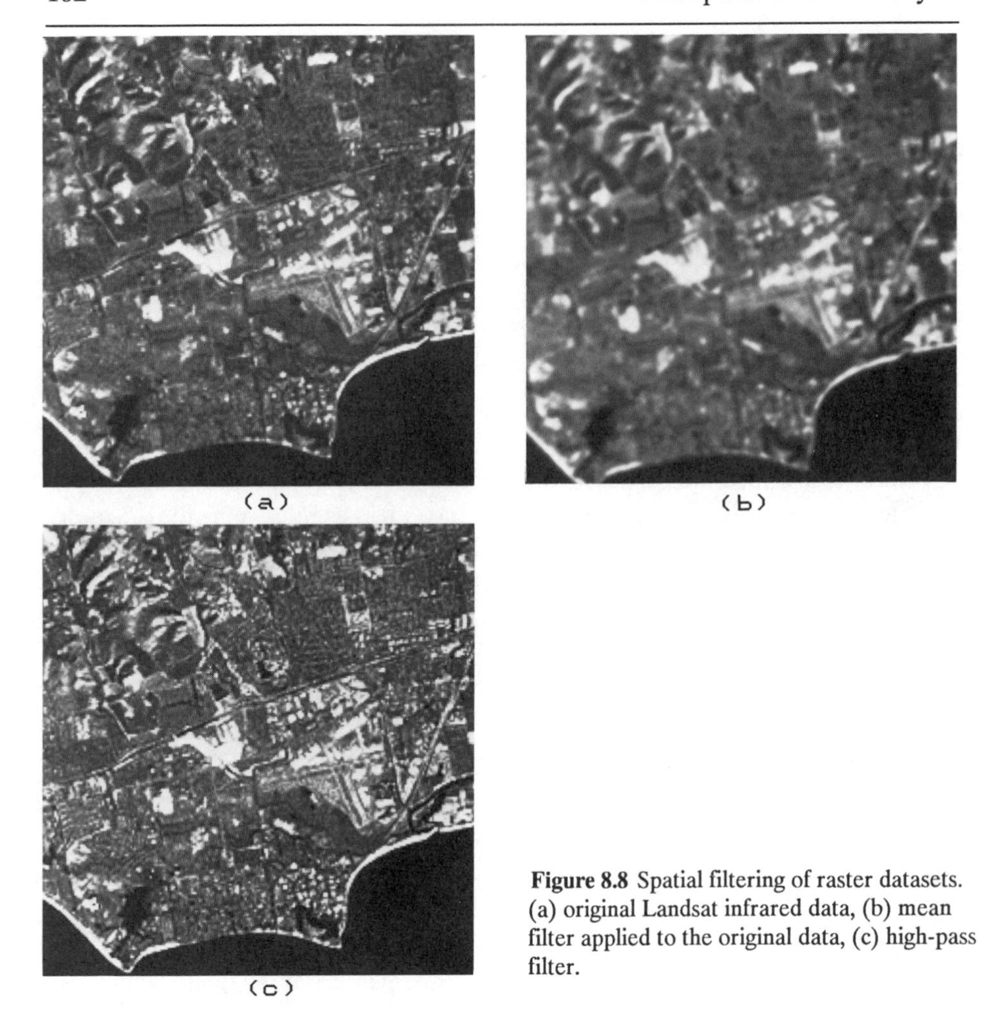

Figure 8.8 Spatial filtering of raster datasets. (a) original Landsat infrared data, (b) mean filter applied to the original data, (c) high-pass filter.

delimited areas, since texture within a homogeneous area (in the way we have defined texture) must be zero. When the neighborhood for the calculation spans the border between relatively homogeneous areas, the texture value is high.

There are a number of spatial transformations that are particularly useful when working with elevation data. A **slope** transformation (Figure 8.9b) turns a data layer of elevation into one of slope, by calculating the local first derivative. As we have seen in the use of spatial filters, many raster-based systems use a local 3-by-3-cell region to calculate the slope at the center of the region. This operation is the same as calculating a local derivative: imagine using the 3-by-3 local set of elevations to determine a best-fitting plane, and then calculate the slope of that plane.

A companion to slope when working with elevation data is **aspect**. An aspect calculation is used to determine the direction that a slope faces (Figure 8.9c). In mountainous terrain, with a ridge oriented east-west, the slopes on the north of the ridge face to the north, and those south of the ridge face to the south. A mathematical way of explaining aspect is to calculate the horizontal component of the vector perpendicular to the surface. Aspect is usually classified into bins of fixed size, so that the resulting data layer is not continuous, but ordinal. A frequent choice is to classify slope into eight categories, each representing an eighth-of-a-circle (or 45-degree) range in aspect.

Another neighborhood operation can be used to determine the characteristics of a dataset along a specified line. Such a process is commonly

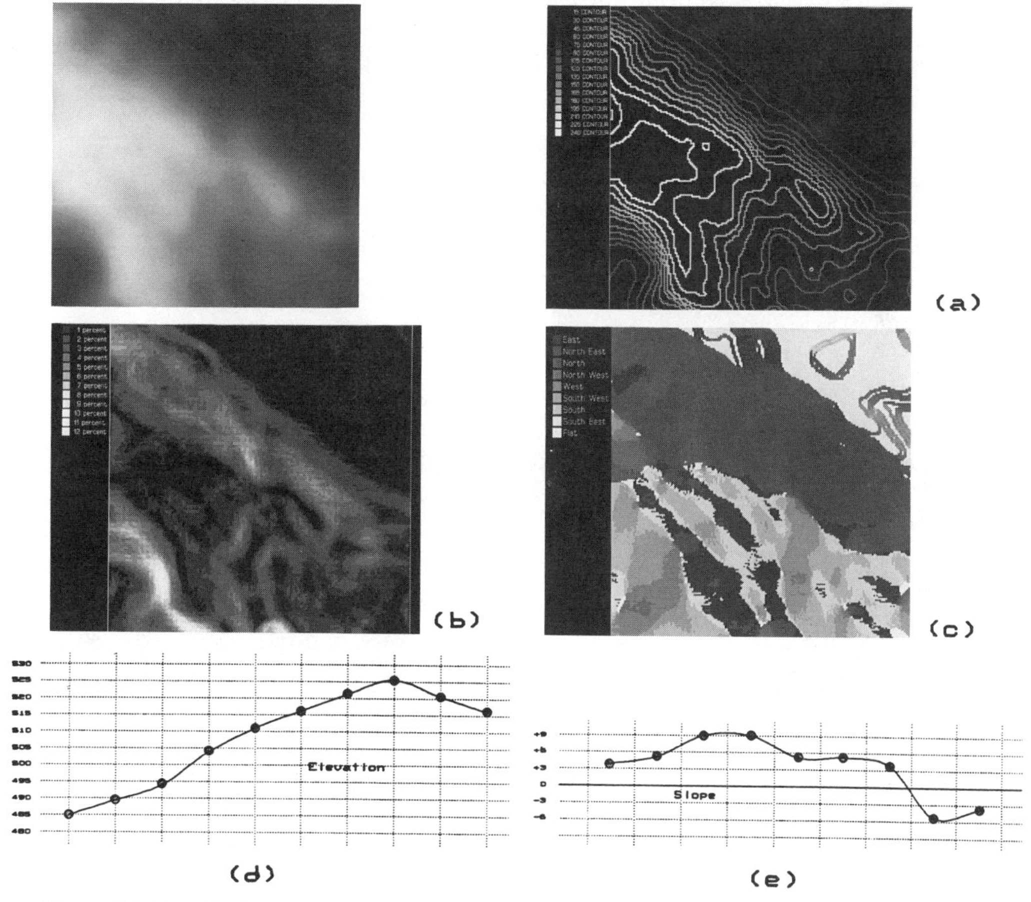

Figure 8.9 Terrain data manipulation. (a) digital terrain data, (b) slope data layer, (c) aspect data layer, (d) elevations along a specified path, (e) slope along a specified path.

graphed as a profile or **cross-section** of a dataset. For example, in the terrain data shown in Figure 8.9, we might need information about the slope of a specified path through the terrain, which corresponds to a ski slope or roadway. There are a number of ways to develop this information, generally based on extracting the elevation values in the neighborhood of the line (Figure 8.9d), interpolating values along the specified path, and then calculating the local slope (Figure 8.9e). Such a calculation may be simpler with raster datasets than with vector forms, at least conceptually. In a raster of elevations, we can use the pixel values near the desired line to estimate elevations along the line, and then calculate the desired set of slopes. In a vector dataset, elevation values are often stored as contour lines. In this second instance, one must effectively fit a smooth surface to the known elevations along the contour lines, and then estimate values along the desired path from the surface.

There are many uses for determination of **visibility** from a point. Impact assessment procedures often consider the visual effects of a new construction project. Tactical planning of military exercises frequently considers sites in terms of their visibility. Calculation of a viewshed -- the terrain that is visible from a specified point -- is a complex calculation, and is similar to one frequently found in computer-aided design systems.

8.6 Measurement

A wide variety of measurement tools are needed in a geographic information system. With raster-based data, the precision of a measurement is of course limited to the raster cell size; vector systems are ultimately limited by the precision of the stored node locations.

Distance measurements are of value in many circumstances. Software components must be available to determine line and arc lengths and point-to-point distance calculations, as well as the perimeter of a specified area. With a vector data structure, the specified region is often a polygon (or based upon the union and intersection of a set of polygons), and the area can be determined from the coordinates of the appropriate nodes. With a raster data structure, polygonal areas are implicit in the data, as collections of locally connected pixels that share a specified set of attribute values. Thus, an operation such as finding the perimeter of an area may require (1) determining the boundaries of the region of interest by searching through the data for changes in attribute values, and (2) totaling the length of the discovered boundary. When calculating the distance between areas, whether the data is based on raster or vector data models, there are at least two different interpretations of distance. The required distance may be at the closest approach between the two polygons, or between a representative location for each polygon (such as the centroid, as in section 8.3). The former may be appropriate for planning an irrigation channel between a pond and a field, while the latter may be better for estimating average travel time between cities.

In Figure 8.10, we see another hole in the golf course we discussed in Chapter 1. A simple distance calculation for this application would be to find the locations for the markers on the sides of the fairway which indicate a distance of 100 yards from the hole. The marker locations could be determined in any of several ways. One sequence of steps might be:

1. identifying the specific hole in the database,
2. use a proximity operator to determine the circle of diameter 100 yards centered at the hole, and
3. locate the markers where the sides of the fairway intersect the circle.

In a raster database, distances between pixels are ordinarily calculated between the center points of the pixels. In a vector database, there are several options. While the distance between two points is unambiguous, the distance between two lines or two polygons, as examples, are not. The distance between two polygons could be calculated as the *shortest* distance between the two, which might be appropriate when considering the most cost-effective location for a canal (along the line of shortest distance) between two bodies of water. One alternative is to calculate the distance between the polygons as the *distance between their centroids*, which might be a reasonable model for the average distance between two parcels.

Area calculations can work in a similar fashion. For example, determining the area of a catchment basin, or the total area of specified land-use zoning in an area, may be required for an application. Thus, returning to Figure 8.10 and our planned golf course, we will need to estimate the area of the green to know how much grass seed and fertilizer to order. In three dimensions, volume calculations may be required. Such applications would include an earth-moving application in civil engineering, or the calculation of the volume of a subsurface water body in hydrology. In our golf course, we could estimate the volume of soil to remove to create the sand trap, or the volume to move to create a level green.

Determination of direction has a wide range of uses. The regions downwind of a smokestack must be identified in an air pollution study. Areas down slope of a water body may be at risk of a catastrophic mud slide after an earthquake. Direction calculations also have applications in route planning, in which a determination of the distance and direction between start and end points provides the necessary navigation instructions.

Another form of measurement involves counting specified objects in a region. If the objects are specified explicitly in the database, the problem is reduced to one of sorting. The system is essentially programmed to sort through the database to find a set of keywords and record the number of times they are found. Some applications may require only a determination that a specified object is found in an area; one example of the object is sufficient. Other applications require an exhaustive search through the database to determine all the occurrences of the desired class of objects. In a raster dataset, the attribute values of the cells in the region of interest are successively compared to the desired values. When a match is found a counter

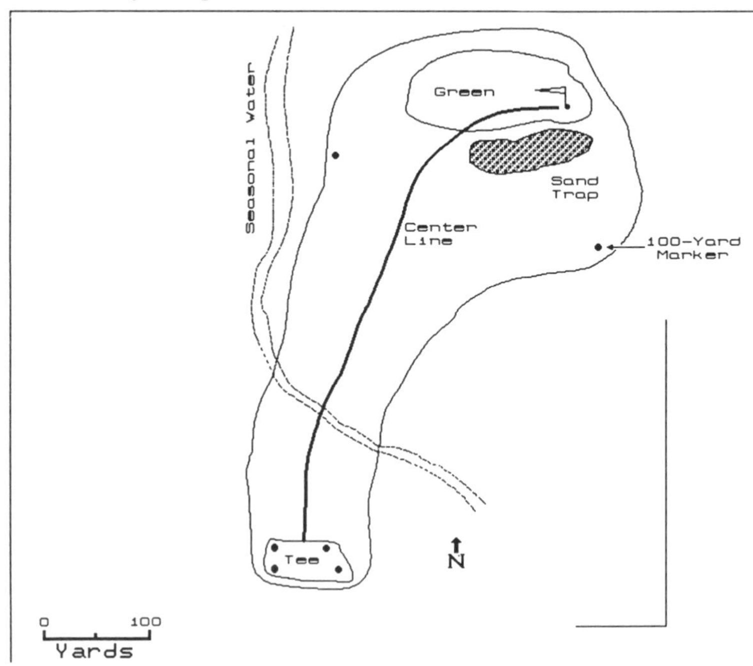

Figure 8.10 Measurements.

is incremented. The procedure is similar for a vector dataset, where the attribute tables are searched for desired sets of values and relationships.

When the desired objects are not explicitly stored in the database, the problem becomes much more complex. Consider the problem of locating a site appropriate for a fire tower in a forest. The problem is mostly that of finding a location with an appropriate viewshed. The data for this problem includes elevation, vegetation cover, and land ownership. There may be algorithmic solutions to such a problem which require a tremendous amount of computation, but there may also be efficient heuristics or "rules of thumb," which permit us to eliminate large areas of the database from consideration. Such strategies enter the realm of artificial intelligence (Smith et al., 1987b).

8.7 Statistical Analysis

There are a variety of statistical techniques that are common in modern geographic information systems. As with many of the other tools we have discussed, these have value in many places in the overall information flow in a GIS. Whether for quality assurance during preprocessing, or for summarizing a dataset as a data management report, or for deriving new data during analysis, several statistical procedures are commonly required. These include:

Descriptive statistics:

The mean, median, and variance of the attribute values in a data layer (or delimited area within a layer) are often needed, for continuous variables. For example, it may be necessary to know the mean elevation of a specified area, or the variance in vegetation density in a field. Higher-order statistical moments, such as the coefficients of skewness and kurtosis which may be used to compare a dataset against a Gaussian distribution, are relatively rare in current systems.

Histograms or frequency counts:

The histogram of a dataset provides us with the distribution of attribute values in a region (section 9.1.3). Note that this calculation is straightforward in a raster, since each cell stores attribute information for a predetermined area; thus, we calculate an area-weighted estimate. On the other hand, in a vector database we have the option of using the area of each polygon to appropriately weight the attribute, or alternately, base the histogram on a *per polygon* analysis. Histograms and frequency counts can be extremely valuable as data screening tools and can help us to formulate hypotheses during analysis. Histograms are useful for examining all types of variables.

Extreme values:

Locating the maximum and minimum attribute values in a specified area is often useful. As an example, in a dataset of bathymetry we may need to find the shallowest and deepest values of a water body. This information is vital during preprocessing, to ensure that the values being entered into a database are reasonable; they are a rapid but not exhaustive way to detect some kinds of coding and transcription errors.

Correlations and cross-tabulation:

In a correlation analysis, we wish to compare the spatial distribution of attributes in two (or more) data layers, usually by calculating a correlation coefficient or linear regression equation when working with interval or ratio variables. For many kinds of data, it is desirable to be able to specify a non-parametric regression; this is rare in modern GISs. For nominal or ordinal variables, a cross-tabulation compares the attributes in two data layers by determining the joint distribution of attributes. When working with both categorical and continuous variables at the same time, the appropriate statistical model is generally an analysis of variance (or analysis of covariance).

We will briefly examine a cross-tabulation, since it is such a powerful and common tool in a GIS. The results of a cross-tabulation are usually displayed in a two-dimensional table:

Household Average per capita income

	< $5,000	<$12,500	<$22,500	>$22,500
Owner	154	354	673	982
Renter	269	627	513	451

The table shown above describes two data layers of demographic information, one regarding per capita income for each household, and the other indicating home ownership. Note that the table is based on categorical data: household ownership is nominal, and per capita income, which is ultimately a ratio variable, has been broken into categories and thus appears in this analysis as an ordinal variable. The data, traditional for census data, may have come from the attribute tables in a relational database model, or from separate income and ownership layers in a raster database model. This specific analysis can be used to examine whether there is a relationship between the level of income and the probability of home ownership. For an analysis of this kind to be complete, there are standard statistical tests that may be applied to determine whether the arrangement of data in the cells of the table might have arisen by chance. Having such statistical procedures within the GIS is very useful. In this cross-tabulation, notice that we have turned one continuous ratio variable plus a nominal variable into an integer-valued ratio variable: this kind of variable type conversion happens frequently.

There will be times when the statistical capability of a geographic information system is simply inadequate for an analysis problem. When this occurs, it will be important to be able to create an intermediate output file, so that data may be transferred to an appropriate system -- perhaps one of the well-known statistical analysis packages (e.g., SPSS, BIOMED, MINITAB). The ability to extract data from the GIS in a standard (or at least, predefined) format is an important consideration for several reasons. In this case, we focus on the need to gain access to statistical software that is external to the system. In section 8.8, we discuss this problem in more detail.

Manipulation and analysis techniques such as these operate on data retrieved from the spatial database, and include both spatial and non-spatial information. New data files containing the "value added" or derived data/information may be created for inclusion within the database. These data can then be analyzed at a later date.

8.8 Modeling

One cannot reasonably expect the developer of a geographic information system to anticipate everything. First, as a user develops experience with a system, the user's interest in complex analytic functions tends to grow. Second, there will be analytic models that are unique to a discipline, or unique to an organization. Perhaps the details of an analytic procedure for an application are proprietary to the company. In these and other instances, we must have means of getting data out of the GIS (as we've mentioned in the discussion of external statistics packages), and into other numerical models for analysis. Furthermore, we should have a means of getting the results of these external systems back into the GIS, for display as well as for storage and further manipulation.

A few common modeling functions should be found in modern systems. The Boolean and arithmetic functions discussed in section 8.1.1 can be used to develop a wide range of applications models. These are of such broad use and applicability that one should expect them in all systems. When it is possible for a user to specify a numerical function by effectively providing a subroutine in an appropriate high-level computer language, the system gains tremendous flexibility and potential for expansion.

A number of considerations bear on the problem of extracting data for use in other software systems. A key component of the problem is the ability to extract a specified subset of the database. Tools should be available to delimit the data of interest both spatially (perhaps by defining a spatial window of interest) and by the desired attributes. Another key component is the ability to determine the specific organization and format for the exported data. Raster data formats, as we have discussed, cover a relatively small range of possibilities, and as such, most geographic information systems possess functions for exporting a univariate raster. Providing functions to specify the raster cell size, and the value coding (for example, floating point format or word size for unsigned integers) are important keys to flexibility. Since vector forms vary more widely, it is more difficult to develop means to export vector files in multiple formats that are widely useful. When such functions are available, it is of course extremely important to have accurate documentation about the detailed formats of the exported data, as well as any available options.

There will always be users who are forced to develop their own software for specialized kinds of analysis. When the internal data formats of a GIS are clearly specified, and when they are appropriate to the analysis problem, it may be possible to write code that operates directly on the GIS database. While this may seem particularly efficient, it opens a path for a number of serious data management problems, including database corruption, lack of appropriate backup protection, and loss of currency. When direct access to the GIS database is desirable, however, we suggest that a reasonable compromise is to be able to directly read the GIS internal database, but not write to it. In this way, the derived data layers can be examined in a separate procedure, and only after appropriate tests have been passed, is the output incorporated into the general database. This is less of a concern if the analytic software is able to "see" only a user's local copy of the database, rather than the overall multi-user database.

There are a number of popular spatial models that are found on some systems. These include determination of visibility or viewshed, network analysis models such as for water flow through piping systems or for calculating optimal paths through transportation networks, and classification and corrections for atmospheric transmission for remote sensing data. In the rest of this section, we briefly describe two applications of modeling in a geographic information system; more are found in Chapter 12.

Spanner et al. (1982) discuss an interesting use of a geographic information system, in which they examine potential soil loss in an area in southern California. According to research they cite, the predicted loss of soil at a particular place, in tons per acre per year, may be estimated as a multiplicative function of rainfall, soil erodability, slope gradient and length, and crop practices. Soil erodability as a function of location was estimated

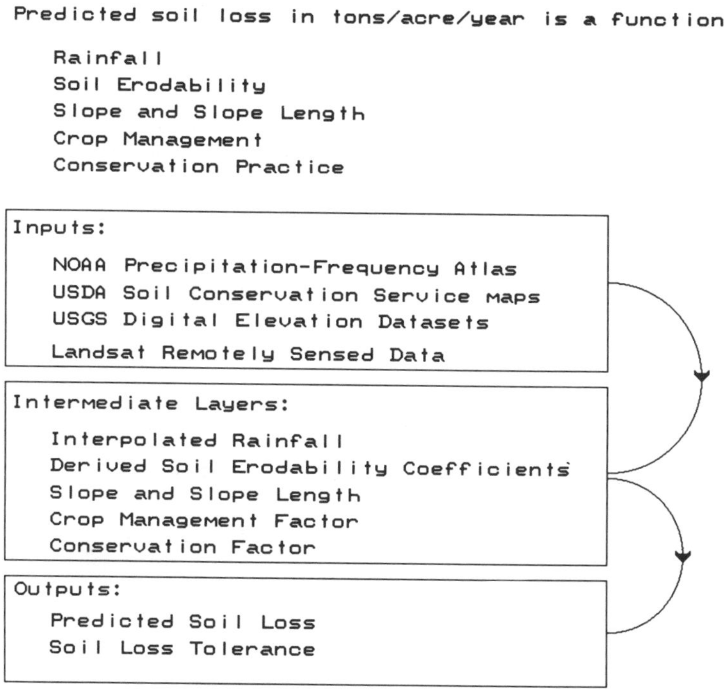

Predicted soil loss in tons/acre/year is a function of:

Rainfall
Soil Erodability
Slope and Slope Length
Crop Management
Conservation Practice

Inputs:

NOAA Precipitation-Frequency Atlas
USDA Soil Conservation Service maps
USGS Digital Elevation Datasets
Landsat Remotely Sensed Data

Intermediate Layers:

Interpolated Rainfall
Derived Soil Erodability Coefficients
Slope and Slope Length
Crop Management Factor
Conservation Factor

Outputs:

Predicted Soil Loss
Soil Loss Tolerance

Figure 8.11 Components of the universal soil-loss equation.

from data collected by the U.S. Department of Agriculture and presented in the form of a map, which was digitized. Length of slope and slope gradient were calculated from a digital elevation model.

The effect of crop management was determined in two phases. First, a map of crop types was developed from a combination of satellite multispectral data, photography, elevation, and site observations. The crop management factor was then determined from guidelines developed by the U.S. Soil Conservation Service. The Soil Conservation Service has also developed estimates of soil-loss tolerance for the soils in the study area; these are estimates that reflect the amount of permissible erosion.

The separate data layers were developed and rectified to a common 60-meter raster grid system. Estimates of soil loss and soil-loss tolerance were then calculated on a cell-by-cell basis. By then subtracting the soil-loss values from the soil-loss tolerance, a final thematic layer is developed that shows those locations where estimated erosion may be a serious obstacle to farming (Figure 8.11). An analysis of the accuracy of the individual data layers, as well as a brief sensitivity analysis of the components of the model, concludes their discussion.

Most geographic information systems have the components needed to estimate a simple cost-of-construction model as a function of location. Consider an organization that needs a site for a new warehouse. The costs of developing the site include building the access road, bringing in utilities (such as power, water, sanitary sewer, and communications), acquiring the land, and building the warehouse itself. Where in the area is the least-cost location for this facility?

The key to such a model is to be able to mathematically describe the component costs as a simple function of location. Consider one component of the problem: building the access road. In the general area of the proposed warehouse, we can identify the principal streets. For any possible location in the general area, we can determine the distance to the nearest street. We can then multiply this distance by the specific cost of constructing the access road (as cost per unit of distance), and thus have an estimate of cost of building the access road to any specified location. In a geographic information system, we generally have proximity operators (section 8.5) that can determine distances from specified classes of features. The input data for this operation is the relevant street network. In a raster-based system, we can develop a data layer in which each pixel is labeled by the distance to the road (Figure 8.12a). In a vector-based system, we might calculate isolines of distance from the outer

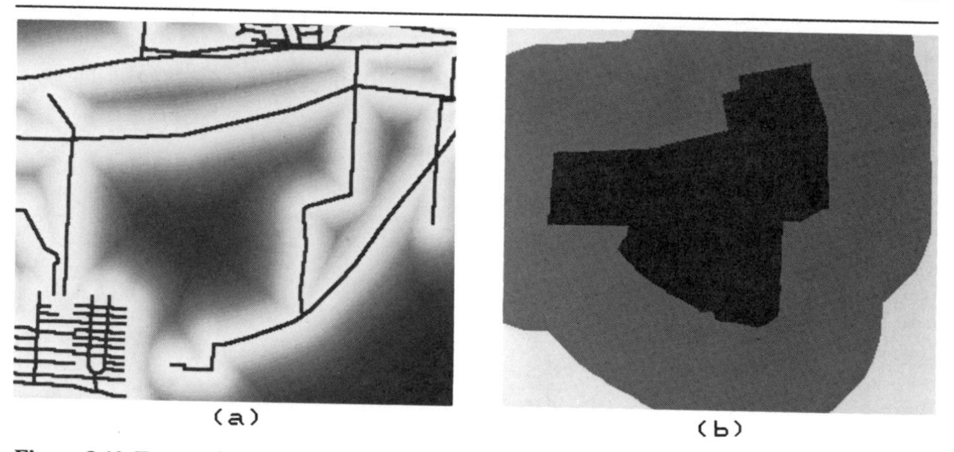

Figure 8.12 Economic cost of construction. (a) proximity on road location, (b) 1 km cost boundary.

margins of the streets, which convey basically the same information. In either case, the distance measures can then be weighted by the cost per unit distance, thus developing one of the component cost layers in this analysis.

In exactly similar ways, we can develop data layers that contain the costs of running power and communications lines, water, and so forth. The costs of acquiring the site for our example warehouse may depend on the site's location. There is an airport in the target area, which affects the cost of land. Sites within 1 kilometer of the airport property boundary are less expensive than those farther away, since the airport is a source of noise and air pollution (Figure 8.12b). This is another application of a proximity operator. However, in this instance we are developing a boundary at a specified distance (which one might consider a binary data layer: inside versus outside the boundary), rather than a continuous surface of distances. If costs are based on a more complicated function, perhaps reflecting aircraft landing and takeoff patterns, these can be added to the model as well.

Finally, we can add the cost of constructing the building itself, which might depend on geotechnical considerations, such as surface soil characteristics and depth to bedrock. A final data layer calculated as the sum of the component cost layers can then be developed, which will indicate the least expensive site for constructing the new facility.

Chapter 9

Product Generation

A geographic information system must include software for displaying maps, graphs, and tabular information on a variety of output media. Cartographic functions should permit the production of the kinds of maps that clearly depict the spatial distribution of various kinds of phenomena. Several non-map graphic products can also be of value when communicating the results of a spatial analysis to the intended audience. The choice of which type of display to use depends upon a number of factors, including the nature of the data itself (for example, whether the phenomena are discrete categories or continuously varying values), required scale and resolution, hardware and software limitations, and of course, the ultimate audience for the output products. Furthermore, we must also have the ability to create non-graphic products as output from a GIS. These have value for transmitting information between different processing systems, as well as maintaining records over long periods of time.

In the following sections we will describe some of the different kinds of products that may form the output of a geographic information system. We will also discuss some of the equipment needed to create these products, to illustrate the range of components now available.

9.1 Types of Output Products

The most common graphics products produced by geographic information systems are maps of various kinds, and some operational systems have comprehensive facilities that create high-quality presentations. We use a very general definition of a **map** (Monmonier and Schnell, 1988): a two-dimensional scale model of a part of the surface of the Earth. This model is a systematic depiction of the earth, generally using symbols to represent certain objects and phenomena. Maps are an effective way of presenting a great deal of information about objects and the spatial relationships of objects. As this discussion makes clear, this is an area where there is a great deal of overlap between a GIS and a computer cartographic system. In order to generate appropriate kinds of maps for an application, the user's cartographic requirements, including scale, projection, annotation, and labels, must be understood. Software components for the production of some of the other types of graphics we will discuss are less common, though valuable in many circumstances. In addition to these, tabular reports of various kinds as well as digital datasets will be required as ways to transfer information to the users of a spatial data processing system.

9.1.1 Common Thematic Maps

Thematic maps concentrate on the spatial variations of a single phenomenon (e.g., rainfall) or the relationship between phenomena (e.g., different classes of land cover). Thematic maps portray the structure of a given distribution, that is, the character of the whole as consisting of the interrelation of the parts. Thematic maps may be used to characterize a wide variety of phenomena.

Choropleth maps are typically used to communicate the relative magnitudes of continuous variables as they occur within the boundaries of unit areas. These kinds of maps are common for census data, such as population density as it varies by county, or average annual per capita income as it varies by country. In these maps (Figure 9.1a), different tones, colors, and shading patterns are used to convey the different values assigned to each predefined polygonal area. The areas are predefined, and thus the potential boundaries between the displayed regions predefined, in that they represent existing political or other boundaries. When constructing choropleth maps, the tones (or colors or patterns) for indicating the data values within regions must be chosen with care.

Choropleth maps are extremely common geographic information system products. In a vector-based GIS, an analytic process can provide us with an attribute value for each predefined polygon, and we can then instruct the system to shade or color the interior of each polygon appropriately. In a raster-based system, this operation is typically more complex, generally using

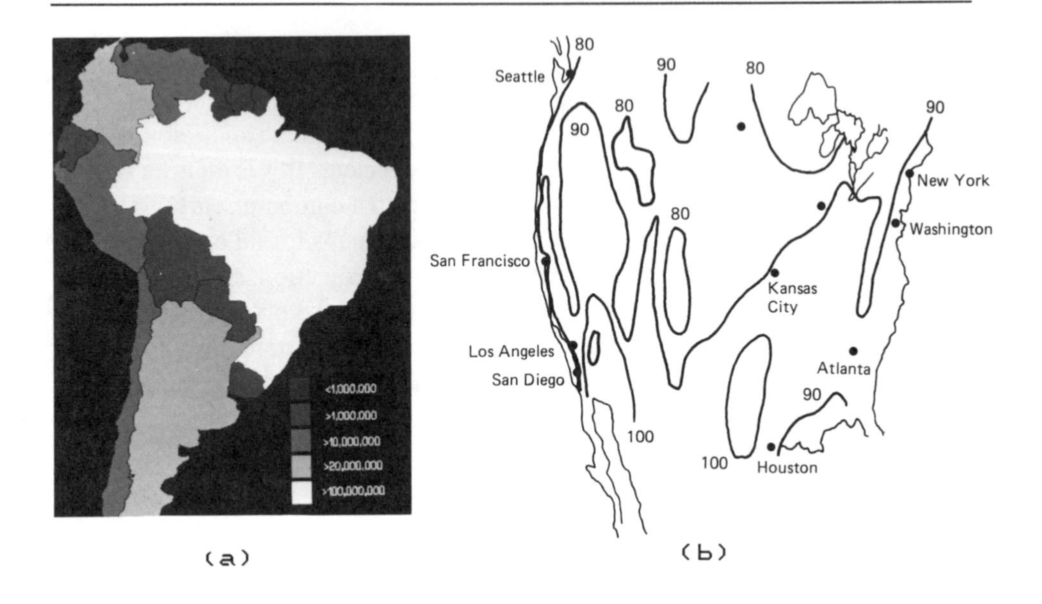

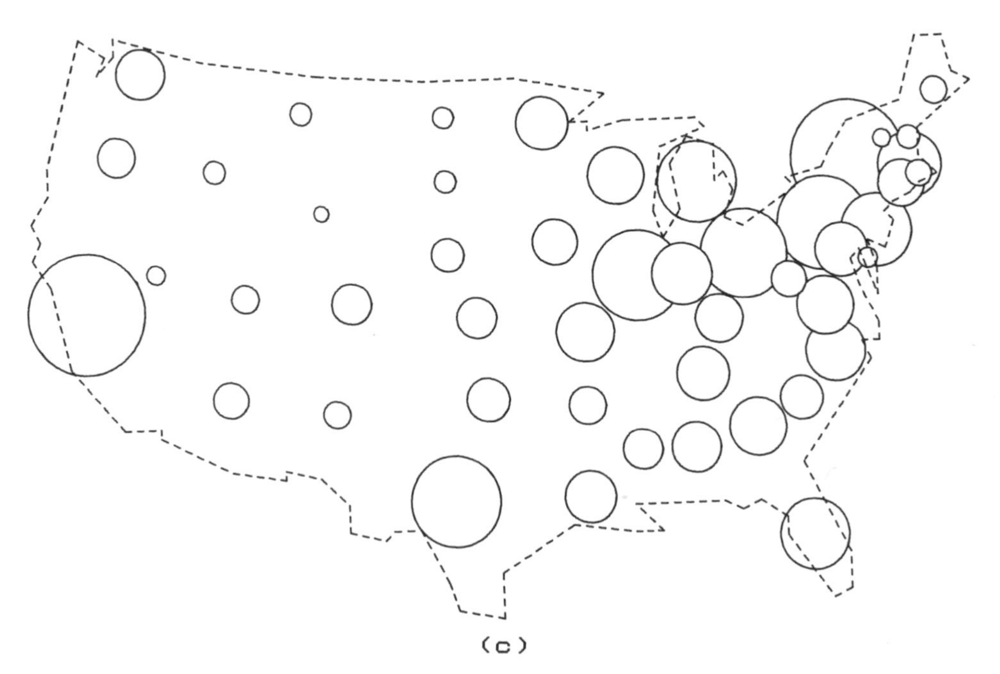

Figure 9.1 Kinds of maps. (a) choropleth, (b) contour, (c) symbol.

a separate data layer each for the area boundaries and the attribute data, followed by a masking process to select the cells for each area and finally by calculation of the area-specific data values; only then is the system ready for the output function. Note that we expect modern systems to be able to include a legend on the displayed choropleth map, so that the correspondence between the data values and the display characteristics is explicit.

The objective of **proximal** or **dasymetric mapping** is to focus on the location and magnitudes of areas exhibiting relative uniformity. Such data as density of crops planted in a region, or land-cover classes, may be represented in dasymetric map form. Like the choropleth maps mentioned above, shading and color patterns are used to describe differences in the thematic values. Unlike the choropleth maps, however, the boundaries are based on changes in the data values themselves, to portray areas that are relatively homogeneous. In a geographic information system, software is almost always available to create maps of this type. Class intervals for the data are created by the analyst, and the system processing is a sorting operation, for determining where the boundaries between classes occur.

Contour or **isarithmic mapping** is used to represent quantities by lines of equal value (which are called the contour lines), and to emphasize gradients among the values. A well-known use of contour mapping is in the display of high and low air pressure regions on a weather map, where the lines represent locations of equal pressure. Perhaps the most common instance of contour mapping is the topographic map (Figure 9.1b), which is based on lines of equal elevation to display the general character of the terrain. A contouring operation is extremely common in a geographic information system. When a contour map must be generated from point source data (such as the information recorded at rainfall gauge stations), the system must first interpolate data values between the known points, and then calculate the positions of the specified contour lines.

On such a map, the choice of the **contour interval** -- the distance between the contour lines -- is crucial to an analyst's ability to understand and convey the patterns in the data. On traditional contour displays (such as a topographic map) the data values are generally not marked for every contour line, but are marked only on the **index contours**. This reduces the clutter on the map and thus improves its overall quality. When the data depicted on a contour map has areas with only small changes in the data values, the contour lines will be spaced far apart. In this case, it is common to add **supplementary contours** at half the usual contour interval. In contrast to this traditional presentation, raster-based systems often use color to code for the different contour-line values. This necessitates the use of a legend to explain the correspondence between color and (in this example) elevation.

Developing a contour map on a raster-based GIS is often done by simply creating an appropriate **look-up table**. The look-up table is generally a portion of the graphic display system, in which the system stores the cross-reference between a cell's attribute value and its displayed representation. Consider a raster of rainfall values, with integer data stored in the raster cells as centimeters per year of precipitation. If we program the system's look up table as follows, we form a contour map with a 4 centimeters/year contour interval:

Cell Value	**Displayed Representation**
1	black
2	black
3	black
4	red
5	black
6	black
7	black
8	red
...	...

The display on the screen would show red contour lines on a black background. In many raster-based systems, the contour lines may be color-coded, so that different colors indicate the different values of each contour line. Note that in a vector-based system, a similar table could be used to show the correspondence between a given contour line's attribute value and its displayed representation either on a plotted map (by selecting an appropriate ink color or pen width) or in terms of color or line type on a television-like display.

9.1.2 Other Kinds of Maps

Dot maps depict spatial distributions by varying numbers of uniform dots. Each dot of a specific size represents the same amount of a given value (such as on a population map of the United States where each dot represents 10,000 people). The software for producing these kinds of maps, and related maps using other symbols, is less frequently found in a geographic information system than in a traditional computer cartographic system. A similar kind of map uses **symbols** of different sizes to indicate values on the map. Figure 9.1c is a symbol map, where the size of circles on the map is chosen to represent the population of a state.

Line maps are used to show the direction and magnitude of potential or actual flows. This is accomplished by the orientation and other characteristics of the lines, such as width or color. An example of such a map would be a depiction of the sources and destinations as well as the volume of products exported from a given country to others (Robinson et al., 1978). These are often of value for network or flow analysis, such as a calculation of stream flow, or tracing patterns of population movement.

Landform maps (Monmonier and Schnell, 1988) are special maps that depict the Earth's surface as if it were viewed from an oblique aerial point of view. Such a synthetic three-dimensional representation requires a significant amount of effort. In section 8.5.2 we discussed some of the other options for presenting this kind of data.

A **cartogram** is another unusual form of map. It is based on the idea that the data display for a homogeneous subdivision of the Earth's surface is not a function of the object's area, but is a function of the value of some other characteristic or attribute of the object. While there are computer cartographic systems which have functions for the creation of cartograms, these are relatively uncommon in geographic information systems.

Animated maps have become more common, as computer graphics techniques make them easier to create. Animation makes it possible to easily display sequences through time; these methods of data display are sometimes called **movie loops**. An example of the use of an animated map would be to show the growth of a city as its population and area increase through time. In this case, the animated map is an excellent summary tool, since patterns occurring through years of change may be efficiently presented to the audience in a minute or two.

9.1.3 Graphics

For some applications and audiences, the results of analyses from a geographic information system may be effectively displayed by means of non-map graphics. Only some of these are commonly available in current-generation GIS's, although they are extremely common in those computer graphics systems that are allied with statistical, financial, and decision support systems.

The overall purpose of graphics is to communicate, to make information easily available to the intended recipients. Wonderful and accessible information on graphic design and presentation may be found in Tufte's *The Visual Display of Quantitative Information* (1983). This volume presents examples (and counter-examples) of quality graphics that were intended to convey information efficiently to the audience. An overall guiding principle of graphic presentation is that a graphic must be able to stand alone: if a great deal of explanation is required to be able to understand the graphic, its design (or construction) has not been successful. We will briefly discuss a few simple and common graphic presentation techniques, which may be valuable tools when working with geographic information systems.

A **bar chart** can be used illustrate differences in an attribute between categories. Bar charts can be organized either vertically or horizontally; in either case, the length of the bar is used to indicate the value of the attribute. In Figure 9.2a, we show a bar chart that portrays the time-varying distribution of land use in an area. Three classes of land use, in each of two years, are displayed. The heights of the bars indicate the total number of hectares covered by each of the land-use categories, and the shading pattern indicates the year of the data. Note the legend in the figure, which explains the correspondence between shading and year. This kind of display can present changes through time in a very clear and effective manner. This graphic might have come from two data layers in a GIS, each recording land use in a different year. To generate these bar charts, a system must be able to accumulate the total area occupied by each of the land-use categories, and then pass the derived data *plus* the annotation to a display module.

A **pie chart** presents information by dividing a circle into sectors, and in this way, illustrates proportions of the whole. In Figure 9.2b, the size of the sector (specifically the area, or to convey the same information, the length of the perimeter) corresponds to the amount of each kind of land use. Note that it is more difficult to determine the actual amount of land use in each of the

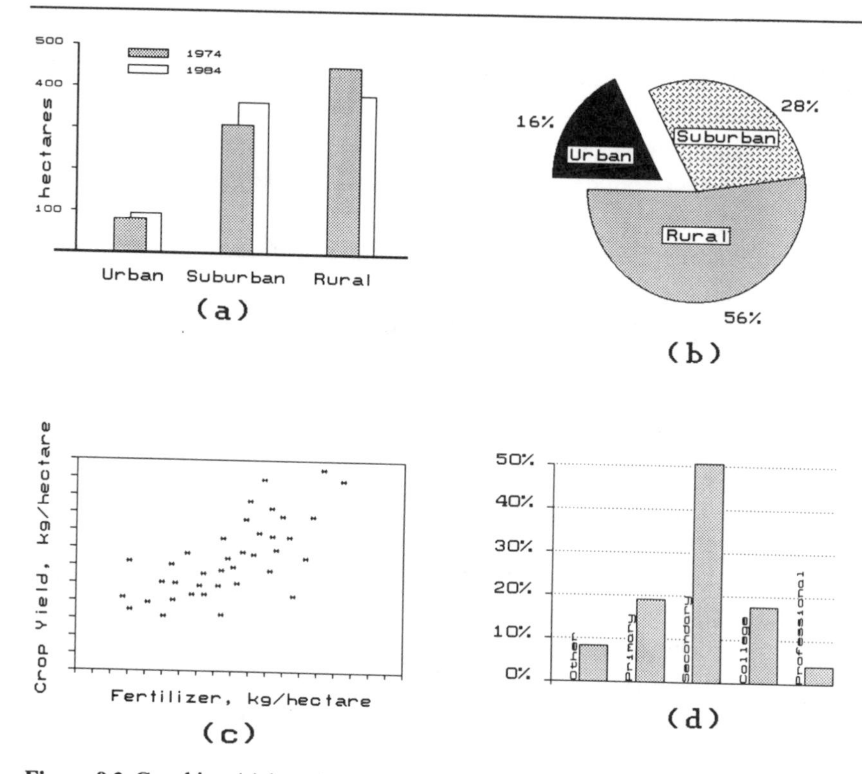

Figure 9.2 Graphics. (a) bar chart, (b) pie chart, (c) scatter plot, (d) histogram.

categories from a pie chart, when compared to the bar chart in Figure 9.2a. For this reason, it is good practice to annotate the pie chart with the fractional coverage of each class, as shown in the figure. Furthermore, one sometimes "explodes" one or more categories from the pie, as we have done here, to highlight a particular category. Pie charts can be very confusing when they must portray a great many categories, since the slices of the pie become small.

Scatter plots are extremely valuable for displaying the behavior of one attribute versus another. In Figure 9.2c, we show a scatter plot of crop yield as a function of the density of applied fertilizer. One can imagine that this data came from a geographic database, and was derived via a complex query. In this case, we have asked the system to compare yield to the amount of applied fertilizer, for those polygons corresponding to fields of a specified crop. To be able to answer such a query, a system would need to select, on the basis of crop type, only the appropriate polygons in an area, and then would extract the two attributes of interest (yield and applied fertilizer) for transmission to the plotting function. Such a display can rapidly suggest many additional kinds of inquiry and hypotheses. For example, in this case there is a suggestion of a curvilinear relationship from the shape of the plot in this figure. The scatter plot is a common capability of many statistics software packages; in more sophisticated systems, various regression lines may be superimposed on the data.

Histograms show the distribution of a single attribute, so that the viewer can examine the way the attribute is apportioned among the different possible values. In Figure 9.2d, in which the data might be from a demographic census of the community, we can easily see what fraction of the population has completed various levels of education. Thus, this presentation displays the frequency distribution of the variable. Such a display is an excellent data screening tool, both to make sure that the data has been recorded correctly (in other words, that all values entered into the database are meaningful), and to evaluate whether the data is appropriate input for certain kinds of statistical analyses (for example, determining whether the data has a Gaussian distribution of values).

There have recently been studies that provide guidelines on the utility and construction of these relatively conventional graphic products. Jarvenpaa and Dickson (1988) provide guidance on, as well as an entry to, this literature.

9.1.4 Numerical Products

Overall, there are two general requirements for numerical products that are frequently used in geographic information systems. First, there should be a reasonable variety of statistical procedures built into the system, as discussed in section 8.7, so that we may produce any of a number of descriptive reports about the data. Second, there must be ways to extract data itself from the system, for any of several purposes we will describe.

Statistical operations on the spatial and attribute data in a GIS cover a tremendous range of possibilities. The elements that must be included in a modern system are those that are useful from a quality-assurance perspective, as well as those that are used frequently for analysis of the data. When appropriate, the results from some statistical operations should be available in graphic form, to effectively summarize and communicate the results. When the desired numerical operations are not available in the GIS, there must be techniques supplied by the system designers for extracting the relevant data granules and passing them to other programs. The same comments can be made in terms of numerical modeling tools.

Another important purpose of a data-extraction capability is to move this data either to another processing/display system, or to a data archive for permanent storage. Some of the details of this concept are discussed in section 8.8. In our laboratory, for example, we have access to several geographic data processing systems, each with different strengths and weaknesses. For many projects, we begin our work on one system, and then must transfer the original and derived data to another for further processing. One of our systems is well-suited for extracting information from maps, another has functions for processing multispectral satellite imagery, another has artificial intelligence tools for heuristic-guided search, another has excellent computer cartographic facilities, and yet another has an expensive high-quality film output device. In this kind of environment, it is important to have tools for extracting portions of a database in one system, and moving it relatively intact into another.

Sometimes, the data must finally be moved to a system where archival storage capabilities exist (for example, optical disk or microform production hardware). In another operational environment, there may be GIS capabilities at different locations in an organization. For example, a central facility may be dedicated to bulk processing of large datasets; this central facility could then extract smaller volume data products for field offices. A similar scenario might be cost-effective where preprocessing tasks need large computational resources but day-to-day operations need only simple query and retrieval algorithms. Finally, networked data systems are becoming the normal data-processing architecture for many organizations. In any of these environments, both original and derived data of many kinds must be able to move between the different locations.

9.2 Hardware Components

Options for displaying material in a geographic information system database should include both **hard copy** display facilities (e.g., paper or film products of various types, including photographic products) and hardware devices for the production of **soft copy** or temporary graphics. Hardware devices for output displays, for both hard and soft copy, include character and graphic printers, plotters of various kinds (pen and electrostatic), color and gray-scale cathode ray tubes, film recorders, and computer output microfilm devices.

As in the discussions of data structures in earlier chapters, we can make a distinction between raster and vector output devices. In order to take advantage of each, we will need the tools we discussed in Chapter 6 to be able to convert from the internal data structure of our GIS, to that required by our output devices.

9.2.1 Raster Components

Typewriter-like printers and commercial televisions are the most familiar raster display devices. With a typewriter, any of a specified number of characters may be displayed, but the position of each character is restricted to a rectangular array of locations. Most inexpensive computer printers work in this way. With such devices, the correspondence between a given attribute value and the printed or displayed character must be determined, and then (hopefully) displayed in a legend.

When working with computer printers, there are two different ways to develop graphic images. In one, characters are selected to represent different categories. For example, a blank character space could code for absence of housing, a solidus (/) could code for low density housing, and the letter *W* could represent high-density housing. In this example we have carefully selected characters whose ink density, and thus, their perceived weight on the page, corresponds naturally to the ordering of the thematic categories of information. In the exact same way, we can choose a sequence of characters whose density on the printed page roughly corresponds with the pixel appearance on the image display monitor (Figure 9.3a).

An alternative is to work with printers in which we have programmatic control over the printing mechanism. With such devices we are able to place dots on the page at a very large number of positions. Popular commercial printers with this capability often can place dots in a rectangular array on the page, generally with an addressable resolution of between 70 and 300 points per inch. In this way, we can develop a set of decision rules on the choice and extent of dot patterns for representing any map or image. Often, areas of constant value in the image are represented by a previously defined pattern of dots that may form different gray levels on the page, or different shading patterns.

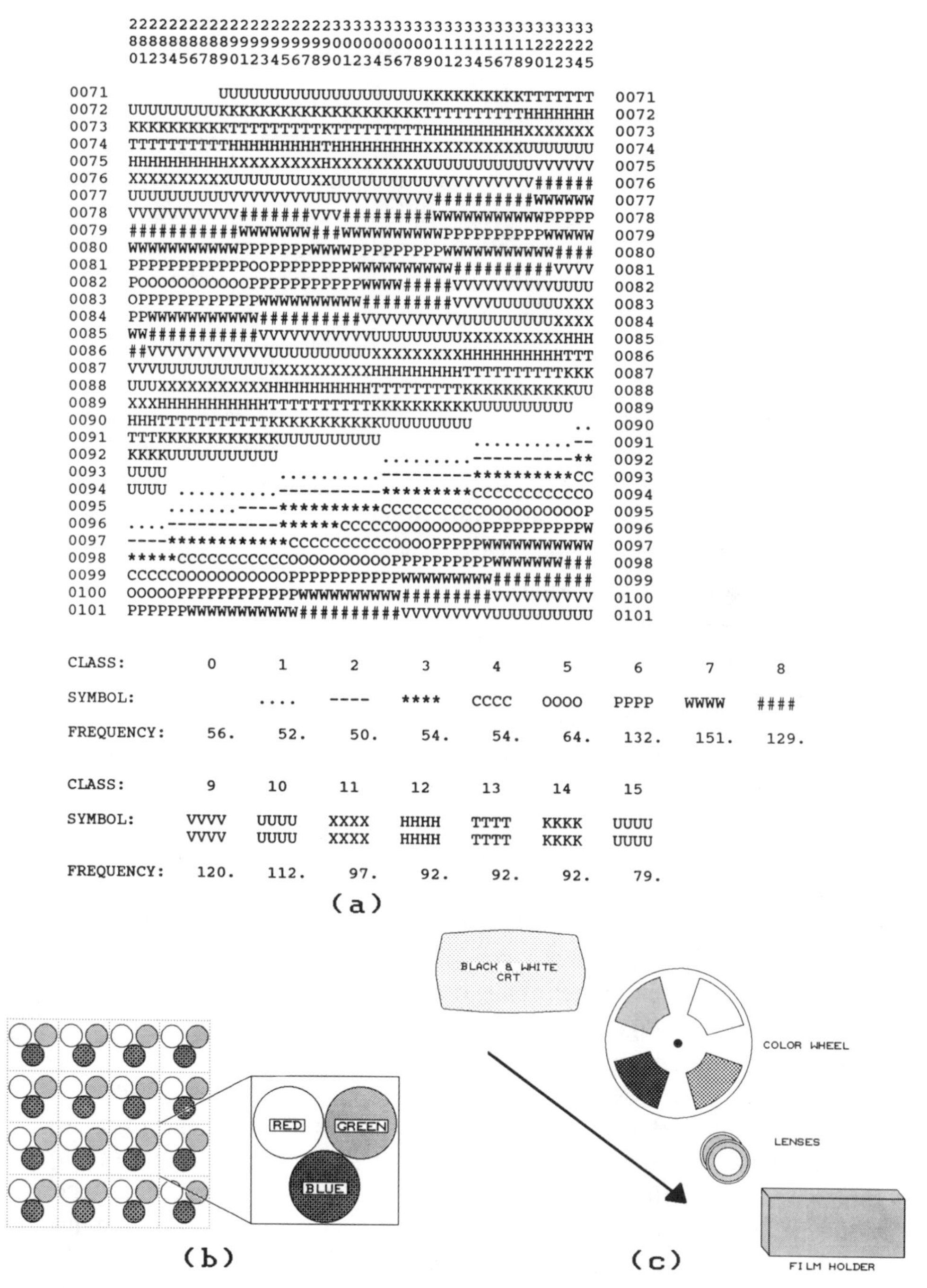

Figure 9.3 Raster display devices. (a) character map from printer, (b) color TV raster display, (c) video film writer.

Similarly, commercial television systems display a rectangular array of elements. On a black-and-white television, each display element on the screen may take a distinct brightness between black and white. On a color television, each display element has three separate dots: one each of red, green, and blue. These triples of color are the smallest elements in the display whose color can be completely specified, and thus, each triple corresponds to a single pixel (Figure 9.3b). The total number of unique colors that may be displayed on such color systems is the product of the number of brightness levels possible for each of the primary colors.

For example, if each of the red, green and blue components may take any of four brightness levels, we can display $4 \cdot 4 \cdot 4$ (or 64 total) colors. When each of the three is as dark as possible, the displayed pixel color is black; when each of the three is as bright as possible, the displayed pixel color is (approximately) white. Since different intensities of gray are displayed when the red, green, and blue components are precisely equal in brightness, we could only display four unique gray values with the system just described. High quality commercial systems often use 256 separate values for each of the red, green, and blue components, and thus, either 256 unique gray levels (when in a monochrome display mode) or approximately 16.7 million potential unique colors. Of course, neither the dynamic range of human visual perception, nor of the display system itself, permits us to discriminate between this many unique colors. However, such systems, which use a full byte for each of the three additive primary colors, are in common use. This kind of system is often called a 24-bit system, since 24 bits, 8 each for the three primary colors, are used to store the color representations. Such displays are often used in electronic color printing and applications of processed satellite imagery.

Displaying horizontal and vertical lines in such raster systems is a simple matter, since such lines correspond to the natural rows and columns of the raster of data itself. Lines at any other angle show a kind of distortion called **aliasing** (see Figure 9.4). The stair-step appearance of the line is caused by the limiting resolution of the raster array, and can be very distracting. In effect, we are periodically sampling the continuous vector at the resolution of the raster cell size, and in this way we are creating a discrete representation of the original data. Raster elements are selected which correspond best to the perfect straight line between end points. For highest performance, an algorithm for selecting these points must avoid multiplication and division, since these relatively complex calculations are slow. One of the best-known implementations of such an algorithm is Bresenham's algorithm (see Pountain, 1987, for a good description). This algorithm is incremental, in that each succeeding step in the calculation is based on the preceding ones. The algorithm computes an error term at each step, which indicates the distance between the raster elements and the ideal line. Points are successively plotted to minimize the error term.

Some systems minimize the stair-step appearance of diagonal lines by employing a technique called **anti-aliasing**. For example, consider that the original vector is to be displayed in white on a black screen, and has a fixed width, irrespective of its direction. As this vector lays across the rows and columns of the raster array, we might record the fraction of each raster cell that is covered by the vector, and then set the brightness of the displayed pixel in proportion to this fractional coverage. In this manner, the transitions between adjacent pixels are less abrupt, and thus the displayed representation of the data is visually more pleasing (see Figure 9.4c). To see this phenomenon, stand this text vertically, and view Figure 9.4 from various distances. The anti-aliased line in 9.4c will have a smoother appearance than the line in 9.4b. This kind of capability is designed into the hardware of many dedicated graphics systems, since it is computationally expensive for a general-purpose computer.

Another family of raster output devices are called **film writers**. These systems are used to make photographic products of many kinds. The simplest

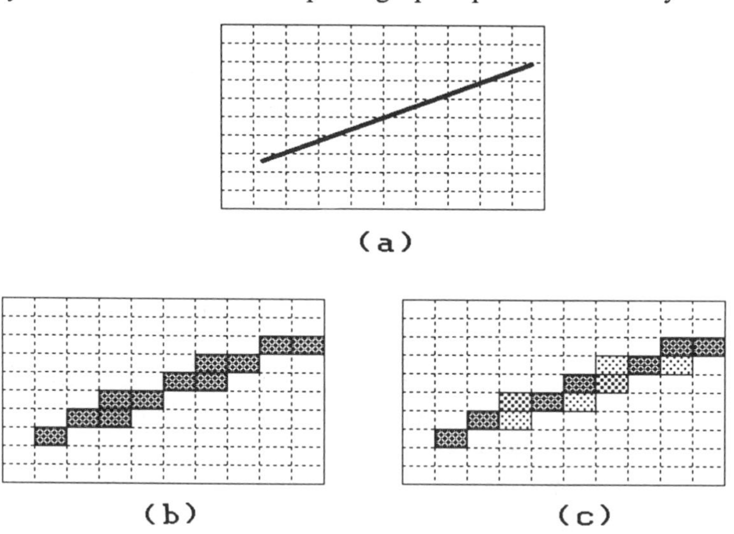

Figure 9.4 Raster distortion. (a) original vector, (b) aliased diagonal line, (c) anti-aliasing applied to the line.

video film writers simply expose film to a monochrome television display. When color is desired, three separate exposures are made on the same piece of color film, one each through red, green, and blue filters (Figure 9.3c). In this way, the triple of color dots used to represent a pixel on a color television display are not visible, since the same monochrome pixel is used three times to develop color. Such systems are capable of images of several thousand pixels on a side. The geometric characteristics of the cathode ray tube and imaging components are extremely important in these systems.

When even more resolution elements are required, electromechanical systems are used to scan a light source (sometimes based on a laser) over the rows and columns of the virtual raster array on the film. These electromechanical devices are much like the film scanners described in section 6.1.2, except that they expose the film to light in a regular pattern, rather than read the film's reflectance.

9.2.2 Vector Components

Vector display devices also come in several forms. A conventional **pen plotter** (Figure 9.5a) operates by moving an arm in two dimensions relative to the paper. Some recent systems operate by moving the pen in one direction and moving the paper in the orthogonal direction. Without going into the details of the system, we may say that the pen can be positioned at any location on the surface of the page. An **electrostatic plotter** simulates a pen plotter, but is based on the technology behind copy machines. A **vector or stroke video display** is similar to this, and very different from commercial television. Unlike a raster video display, the electron beam which draws the image in a vector video display can be moved to any arbitrary location on the screen. With such a vector display, points may be drawn anywhere on the surface of the screen, and there is no apparent aliasing of lines. This ability follows from the underlying analog nature of such systems.

Refresh vector devices use displays that are usually based on a **display list**. The display list stores a current record of all the graphic elements to be displayed. The hardware components continually loop through the display list, writing the appropriate graphics to the screen as they traverse the list. When the display list is too long, the screen display may flicker; the system designers must balance the expected complexity of the graphic image against both the speed of the hardware and the temporal persistence of the display screen

(a)

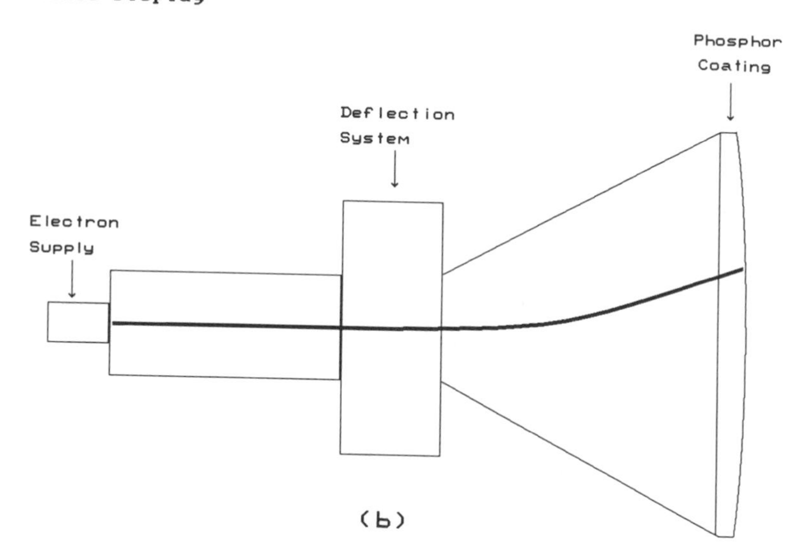

Figure 9.5 Vector display devices. (a) pen plotter, (b) stroke display.

phosphor. **Storage** type vector displays are based on a different technology. In this alternative, graphic elements written to the screen are stored as an image on the display tube itself. As such, the length of the display list doesn't affect the quality of the display; we trade complexity in the display driving hardware and software against complexity in the design of the cathode ray tube itself. Vector video displays are rarely able to display more than a few distinct colors.

The choice of geographic information system output hardware and software is largely dependent on the application. Color may be valuable in some circumstances, photographic materials in others. Large paper products may be necessary in some cases and not in others. As we will see in Chapter 12, GIS systems are being directed at a wide variety of applications, with a concomitant need for a wide variety of output products.

Chapter 10

Remote Sensing and GIS

Remote sensing and image processing are powerful tools for many research and applications areas. **Remote sensing** may be defined as the process of deriving information by means of systems that are not in direct contact with the objects or phenomena of interest. **Image processing** specifically refers to manipulating the raw data produced by remote sensing systems.

Remote sensing often requires other kinds of ancillary data to achieve both its greatest value and the highest levels of accuracy as a data and information production technology. Geographic information systems can provide this capability. They permit the integration of datasets acquired from library, laboratory, and fieldwork with remotely sensed data. On the other hand, applications of GIS are heavily dependent on both the timeliness or currency of the data they contain, as well as the geographic coverage of the database. For a variety of applications, remote sensing, while only one source of potential input to a GIS, can be valuable. Remote sensing can provide timely data at scales appropriate to a variety of applications. As such, many researchers feel that the use of geographic information systems and remote sensing can lead to important advances in research and operational applications. Merging these two technologies can result in a tremendous increase in information for many kinds of users.

Remote sensing is a technology that has close ties to geographic information systems. Aerial photographs may be interpreted by trained analysts, who then record their interpretations in an overlay containing the locations of various features. In a subsequent step, the overlay may be traced on a digitizing tablet for input into a GIS. In another application, a multispectral scanner dataset may be processed to extract a selected set of land-cover classes using machine-assisted techniques. In both of these cases, remote sensing represents a powerful technology for providing input data for measurement, mapping, monitoring, and modeling within a GIS context. Indeed, it has been suggested that neither remote sensing of Earth resources nor geographic information systems can reach their full potential unless the two technologies are fundamentally linked (Estes et al., 1984). This is very likely the case as geographic information systems, to be most useful, must have the up-to-date information that can often be extracted from remotely sensed data.

In a complementary fashion, image processing and interpretation of remotely sensed data must employ ancillary or collateral data (such as elevation data and existing land-use/land-cover data) to achieve high levels of thematic classification accuracy, which naturally brings us to work in the GIS environment. The material that follows provides a brief history of the field of remote sensing, discusses some essential elements of the field that provide important links to GIS technology, and briefly examines a number of existing aircraft and satellite data products that can be used as input to a GIS.

From a philosophical point of view, applications of remote sensing are fundamentally in the realm of geographic information system, since the same concerns, and the same overall processing flows, are found. Seen this way, processing systems for remotely sensed data may, in large measure, be considered a specialized form of GIS. Remotely sensed data does, however, have certain unusual features, which may necessitate special-purpose components in the system. Furthermore, digital remote sensing systems often create immense volumes of raster data, which may stretch current computer processing and storage systems to their limits. Overall, however, we view a GIS as a generic system for manipulating spatial data, and see remote sensing and image processing as more specialized techniques of such systems.

The launch of the Landsat satellite system in 1972 began a period of major advances in the science and technology of remote sensing, as well as in the related disciplines that use these tools. Traditionally, field studies in land sciences have been limited in scope. Such studies often involve only one or two variables and are focused within a relatively small area. This was due in part to our inability to acquire, manage, interpret and manipulate large volumes of heterogeneous data. This inability has restricted our ability to understand global problems. Modern remote sensing technology provides a number of efficient and consistent ways to gather spatial data. Indeed, remote sensing offers the earth sciences and applications community globally consistent datasets for the first time.

An excellent compendium of detailed information about remote sensing systems, theory, and practice is the *Manual of Remote Sensing* (Colwell, 1983). This two-volume set has useful information for generalists and specialists alike in its 2400-plus pages.

10.1 A Brief History

The *Manual of Remote Sensing* subdivides the development of remote sensing technology and practice into two time periods. Prior to 1960, for all practical purposes, aerial photography was the sole system used in remote sensing. With the commencement of space programs in the early 1960s, the pace of technological development quickened, the range of sensor systems expanded, and the amount of digital remotely sensed data available increased. Figure 10.1 provides a context for the developments which have occurred in remote sensing since the early 1960s. The development of camera systems began some 2,300 years ago when Aristotle experimented with the principle of the *camera obscura*. This attempt to permanently record images set the stage for a series of experiments that would continue up through the nineteenth century, culminating in 1839. In this year Louis Daguerre announced the **daguerrotype**, a photographic process developed by him and Joseph Niepce. This process was refined quickly; optical quality glass was developed, and lens design was incorporated into the development of camera systems. Thus by the turn of the century the foundations of modern photography were in place.

At the same time these developments were occurring, researchers were working to extend the sensitivity of photographic emulsions beyond the limits of the visible portion of the electromagnetic spectrum (see Figure 10.2). These efforts were successful and ultimately led to the development, among other things, of reflectance infrared photographic emulsions, in both black-and-white and color. This was followed by the development, in the period between World War II and 1960, of electro-optical sensor systems that could operate at ultraviolet, visible, reflectance-infrared and thermal-infrared wavelengths. Also during this period, practical side-looking airborne radar (SLAR) systems were developed. Thus, the technology of remote sensing had moved from systems that were largely confined to daylight operation, to those that could operate day or night, and under all but the most severe weather conditions. With these advances in sensor systems, we began to acquire images in analog and digital form rather than solely in analog photography.

1820	Niepce takes first photographs of nature
1859	First aerial photographs - captive ballon over the French countryside
1862	Forest mapping from aerial photographs
1910	Wilbur Wright takes first photographs from an airplane
1920s	Systematic forestry mapping from aerial photography in Canada and the United States
1960	TIROS 1 - first operational meteorological satellite
1962	Prototype multispectral camera constructed by Zaitor and Tsuprun
1962	Mercury 8 astronauts take first photographs of Earth
1966	Digital image analysis for agricultural applications
1972	Launch of Landsat 1
1978	Launch of Seasat
1982	Launch of Landsat 4 - Thematic Mapper sensor
1986	Launch of SPOT

Figure 10.1 Remote sensing timeline. Some of the important dates in civilian remote sensing programs.

Once remote-sensing devices have collected their image data, the images must be interpreted. These representations of the environment can be analyzed either by human interpreters employing what has been termed manual image analysis procedures, or by human interpreters using computer technology. This second type of processing has been variously referred to as computer-assisted or automated image analysis. The latter term, automated image analysis, is a misnomer since human interaction is always required at some stage in the process. Because the image analysis process is an important technique for deriving information for input into a geographic information system, we will discuss the relationship of manual and computer-assisted analysis in more detail below.

Another key element in the increasing importance of remote sensing as a data-acquisition technology is the development and deployment of operational Earth-orbiting satellites capable of recording large amounts of data in digital form. These satellites provide vantage points, synoptic views, and repeat-coverage capabilities that are not possible or practical from aircraft-mounted sensor systems. There are today, however, a wide variety of remote sensor systems capable of providing data for analysis and input into a geographic information system.

10.2 Remote Sensing Technology

Remote sensing can cover a wide range of spatial and temporal scales (see, for example, Colwell 1983, p. 25). We are all familiar with the satellite-derived weather maps in our morning or evening newspapers. The sensing devices for collecting this data cover a large fraction of the Earth's surface on a more or less continuous basis. At the other extreme, some of us remember the first U.S. space shuttle launch, when Earth-bound imaging systems were able to determine that the heat-resistant tiles on the orbiting shuttle were indeed still in place. There are many monographs and professional publications on the technology and applications of remote sensing. These include such journals as *Photogrammetric Engineering and Remote Sensing* and the *International Journal of Remote Sensing*. Our purpose here is to highlight the elements that a professional in an allied area might need to know.

As a starting point, we make a distinction between two kinds of remote sensing. Photographic systems involve cameras, with their lenses, shutters, and energy-sensitive emulsions. Photographs are made when electromagnetic radiation -- usually reflected visible light -- passes through the lens of the camera, strikes an emulsion surface, and forms a latent image. This imaging system is to be distinguished from other imaging systems that use other portions of the electromagnetic spectrum, portions that the human eye cannot detect. Non-photographic two-dimensional image recordings can be made from such electromagnetic radiation sources as infrared, ultraviolet, and radio frequencies, reflected or emitted, by receiving the radiation with a detector or antenna system, converting it to an electronic signal, and recording the signal. Thus, photographs are always a kind of image, while images are only sometimes photographs.

Another important distinction to make is between **passive** and **active** sensing systems. Passive systems detect electromagnetic energy that has been supplied by some other source. A simple camera with natural color film passively detects the energy of the sun, after the energy has been reflected by the target, in a wavelength band of roughly 400 to 700 nanometers. Active systems, on the other hand, provide their own energy source. To cite a familiar example, in a very real sense photographs taken with a flash lamp are the result of an active remote sensing system.

All remote sensing systems are designed to acquire imagery in one or more wavelength bands of the electromagnetic spectrum. Each band of image data acquired will potentially have a different information content, and multiple-band imagery has been used extensively for machine-assisted extraction of information concerning environmental objects and phenomenon. We return to this important area later in this section.

10.2.1 Photographic Sensors

Details of photographic systems and their operation are covered in many textbooks (see, for example, Lillesand and Kiefer, 1987; and Wolfe, 1983) and other sources (Slater, 1983). We briefly outline some of the essential concepts here. In selecting a particular photographic system for some application, there are a number of key decisions that must be made. These decisions concern the following areas:

Film emulsions:

A given film emulsion may be characterized by its spectral response. The spectral response (Figure 10.2) describes the sensitivity of the film to light of different wavelengths. **Natural color** film has a series of chemical layers, with the overall result being that a print or transparency presents a view that is close to that of human vision. **Color infrared** film, on the other hand, ignores light in the blue part of the visible spectrum, and is sensitive to some of the energy in the infrared portion of the spectrum. Vegetation and geological materials each have strong reflectance signatures in the infrared, and consequently, color infrared film has been used to great effect when surveying these materials.

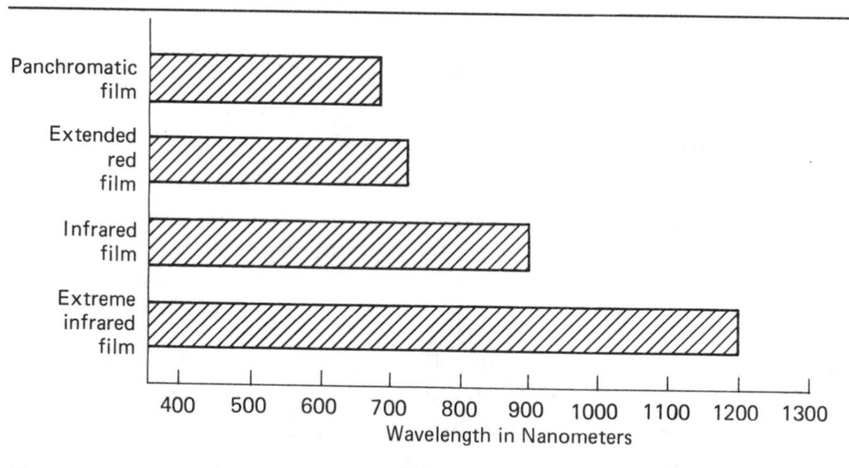

Figure 10.2 Spectral response of film emulsions. The short wavelength cutoff at approximately 360 nm is caused by the spectral transmission of most optical glass.

Lens system:

The camera lens governs the system's ability to collect electromagnetic energy. The **focal length** of the lens describes the magnification of the system. For aerial photographic cameras, focal lengths of 6 and 12 inches are common in the United States. A 12-inch camera lens will magnify objects on the film a factor of 2 (in linear dimension) in comparison to a 6-inch lens. A simple trigonometric relation converts the focal length of the lens and the size of the film frame into the area covered on the ground at a given elevation (see Figure 10.3).

Shutter system:

The camera shutter controls the entry of light onto the film emulsion. A short exposure provides little time for the motion of the camera or the target to smear the image developing on the film. Higher film sensitivity permits shorter exposure times. However, if the exposure time is too short, there will be insufficient interaction between the incoming light and the film, resulting in an unsatisfactory image.

Distance from target:

When the target objects all fall in a flat plane, the film is held parallel to that plane, and the optical axis of the lens is perpendicular to these two, there is a constant relationship between distances on the film and corresponding distances on the object plane (Figure 10.3a). When these conditions are met, a unit of distance on the film will be related to a unit of distance on the ground by a constant scale factor.

Unfortunately, maintaining this relationship with high precision is quite difficult. Tilt in the camera orientation (Figure 10.3b), variations in the terrain (Figure 10.3c), and changes in the platform motion all cause the distance between the objects on the ground and their corresponding locations in the film plane to vary. Removing these variations can be quite complex; we briefly addressed this kind of problem in section 6.8.

There are a series of different kinds of geometric distortions that are due to the systems we depend upon for data. Complex geometrical problems can be found in imaging systems. With the increasing use of space and aircraft imaging sensors to provide digital data to geographic information systems, it is important to understand some of the inherent problems in the use of this kind of data. Many remote sensing systems are based on photographic technology, using film and shutters. Others use a photodetector and a scanning mirror: the scanning mirror provides deflection of the field of view of the detector across the platform's path, which usually corresponds to the rows in the resulting raster image. The platform's forward motion provides movement to

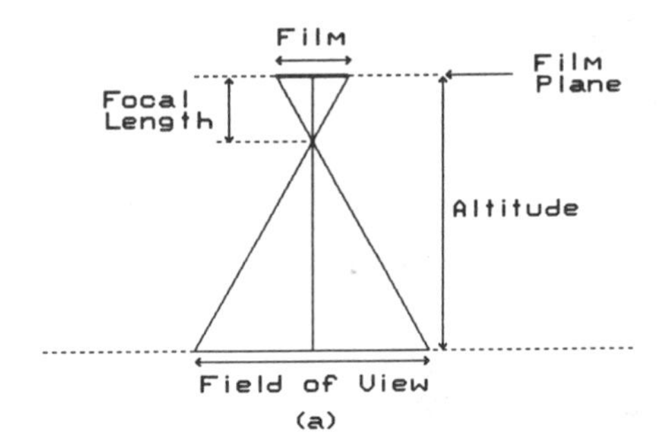

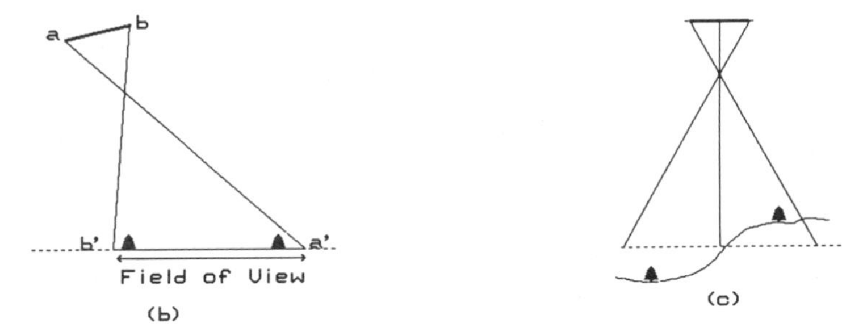

Figure 10.3 Camera viewing geometry. (a) normal orientation, (b) tilted, (c) the effect of non-level terrain.

succeeding rows in the raster array. More recent technologies, based on line or area arrays of detecting elements, do not use a scanning mirror, and thus are not subject to its distortion. However, line and area arrays have their own unique problems, including imperfections in the array geometry, non-functioning elements, and gain and offset differences for each element.

Some of the typical distortions in a sensing system include (Figure 10.4):

Perspective Distortion:

As mentioned in the previous section, scale in a perspective view of the Earth is not constant. Imagine a photograph of the Earth, taken so that the downward (or **nadir**) direction corresponds to the exact center of the photograph (called the **principal point**). An object will appear larger in this image when closer to the principal point than when further away. The reason for this effect is simple: objects closer to the principal point are actually closer to the camera than objects seen in the photograph's margins. The same effects are seen in most non-photographic sensors.

Scan angle distortion:

If the velocity of the rotating scan mirror is constant, the distance between the centers of successive pixels along a row is a function of the mirror deflection from nadir. Thus, pixels far from nadir cover a larger area on the ground than those close to nadir.

Earth's rotation:

If the Earth's rotation is significant during the time a frame is being acquired, the resulting image is skewed. The problem is simple to explain. If the imaging system takes a significant amount of time to acquire all the rows in an image, the Earth will have rotated a significant amount during the acquisition time. Thus, for a sensor in a polar orbit moving from north to south, the upper left corner of the image is **not** due north of the lower left corner, since the Earth has rotated. In other words, the resulting square image corresponds to a area on the ground that is **not** square.

Uncontrolled platform motion:

Roll, pitch, yaw, altitude, and velocity changes during acquisition will all cause distortions of various kinds, including smearing of the image and non-systematic scale changes.

Topographic distortion:

Variations in topography will cause distortions in the image, since changes in topography change the effective ground-to-sensor distance, and thus the scale factor. Ground features at higher elevation are closer to the sensor, and thus appear larger, than identical features at lower elevation. This effect is more important for low-altitude sensors than for high.

Lens aberrations:

All lenses slightly degrade the formed images, and the imperfections are called aberrations. Spherical aberrations degrade the portion of the image near the axis of the lens. Coma is a degradation at the edges of the image. Astigmatism comes about when objects at the same distance from the lens do not focus at the same image plane. Chromatic aberrations are seen when light of different wavelengths is dispersed differently by the lens.

When the distortions present in the sensing system are systematic, rather than random, it may be practical to devise a consistent set of corrections for an imaging system. For example, if more than one photosensor is involved in the imaging system, we can develop equations to compensate for differences

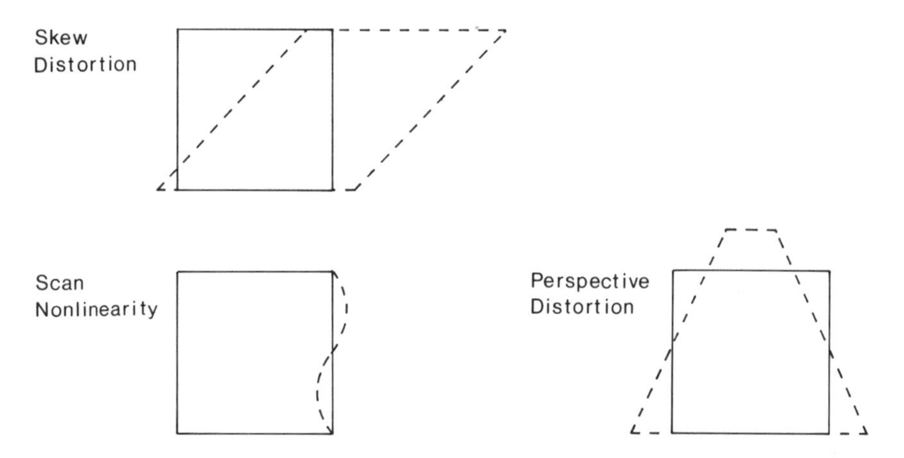

Figure 10.4 Distortions in a sensing system (after Moik, 1984).

in gain and offset between the different photosensors. This is the case in the early Landsat multispectral scanners. If image skew due to the Earth's rotation is constant, we can remove the distortion mathematically.

There is a great deal more to consider in the practical applications of these sensing technologies to spatial data processing. Such practical matters as mission planning, and such technical matters as compensating for the platform's forward motion during exposure, are discussed in some detail in the references mentioned at the start of this section. Photography is still the primary material used in modern topographic mapping, and is very important for such applications as soil surveys, forestry, and urban planning. Overall, the use of aerial photography is at present far more widespread than the use of imagery from non-photographic sensors, which we introduce in the next section. Photographic systems are very powerful tools that record a tremendous amount of information in a small space, often for a relatively small cost. Nevertheless, interpreters must exercise all their skill to make maximum use of the materials (see section 6.8). In addition, conventional darkroom techniques are limited in comparison to the kinds of digital processing we have discussed in Chapters 6 and 8.

10.2.2 Non-Photographic Sensors

We focus our discussion in this section on imaging sensors: those systems that record a two-dimensional array of data (which often may represent an area on the ground). Robinson and DeWitt (1983) present a clear discussion of non-imaging sensors. The *Manual of Remote Sensing* (Colwell, 1983) also provides a tremendous wealth of detailed information about this subject. The following should serve as an introduction.

As in photographic systems, there are several essential characteristics to consider in a non-photographic imaging sensor.

Spectral sensitivity:

A sensor's sensitivity as a function of wavelength forms one of the primary specifications of the system. Sensors in operational use on aircraft and spacecraft platforms have generally from one to twelve discrete spectral channels. Some of the new research instruments have more than 200 discrete spectral bands. We address the value of different spectral bands later in this section.

Spatial characteristics:

The field of view of a sensor, as in the case of a camera system, is normally described either in terms of the length (or width) of the scene as viewed from a nominal operating altitude, or in terms of the angle subtended by a line from the edge of the scene, to the sensor, and back to the center of the scene. These measurements are governed by the optical and geometrical characteristics of the sensor system. In addition, we define the **instantaneous field of view** (IFOV) as the field of view that is observed at an instant in time by a single detector. Many non- photographic systems have multiple detecting elements, and some may sweep these elements over the field of view during an interval of time. Thus, the IFOV is typically much smaller than the overall field of view. In the context of a geographic information system, the IFOV ordinarily corresponds to the size of a single raster cell.

Temporal characteristics:

This involves questions of the time sequence of a system's coverage of a geographic area. With aircraft sensors, repeat coverage is generally at irregular intervals. With satellite sensors, such as Landsat or the Advanced Very High Resolution Radiometer on the Nimbus series of satellites, the potential exists for routine repeat coverage cycles. In addition, the geostationary meteorological satellite systems permit repeat coverage ranging from minutes to weeks, cloud cover permitting.

The discussions in the previous section about changes in the scale of an image as a function of the viewing geometry are completely relevant for non-photographic sensors.

When discussing sensors flown in space, there are a number of considerations that focus on the spacecraft orbit. At one extreme is a **polar orbit**, where the ground track (the locus of points on the Earth's surface directly beneath the craft) of the spacecraft is pole to pole. Note that this does *not* mean that the ground track is a great circle corresponding to a longitude line: while the spacecraft is revolving around the Earth from pole to pole, the Earth is rotating on its axis beneath the spacecraft. At the other extreme is an **equatorial orbit.** A low inclination orbit is one in which the ground track is restricted to a band of latitudes near the equator.

The altitude of the craft is of course another important specification, in part because of the relationship between spacecraft altitude and the time the craft takes to complete an orbit. An orbit is termed **sun-synchronous** if the time at which the craft passes overhead at a location is the same on each overpass. Many operational sensors are flown in sun-synchronous orbits, so that the gross effects of sun angle on the acquired data (such as the creation of shadows) are relatively constant. Some meteorological satellite systems are in **geosynchronous orbit**, which means that the satellite platform remains essentially stationary over the same point on the Earth. In this way, the repeat coverage cycle is dependent only on the sensor system, since the orbit permits continuous observations of the same area. For example, two such geostationary satellites, GOES East and GOES West, image from the Atlantic to the Pacific to observe weather patterns.

There are two common configurations for the detecting elements of sensors operating in the visible-through-infrared portion of the spectrum. The older technology uses a scanning mirror to sweep the instantaneous field of view of a detector across the track of the platform, which may be considered the rows of the resulting raster data array (Figure 10.5a). This overall mechanism is called a **whisk-broom scanner**, since the across-track motion of the detector field of view reminds one of a small broom, in effect sweeping across the surface of the Earth. Forward motion of the platform accomplishes the data scan in the column direction of the image.

In operational systems, there will normally be separate detectors for each spectral band. Furthermore, there may be multiple detectors for each separate band, as in the Landsat multispectral MSS and TM sensors. Additional detectors for a specified wavelength interval may share the scanning system. When this is the case, the across-track scan speed may be lower, since there are several parallel rows being acquired during the sweep interval. Returning to the broom metaphor, there are several separate bristles in the broom. These systems are mechanically complex, produce undesirable vibrations in the sensor package, and permit the detectors only a very limited time to dwell on each effective ground pixel.

An alternative technology, of more recent development, is called a **push-broom scanner**. This is the technology used in the French Système Probatoire de la Observation de la Terre (SPOT). In this case (Figure 10.5b), there is an optical detector for each pixel in the across-track direction, arranged in a line in the sensor (Goetz et al., 1985, presents the details of one experimental instrument). The lens system focuses each ground pixel on its own detector. There are several advantages to this design, in comparison to the whisk-broom system. There are no moving parts, since there is no need for

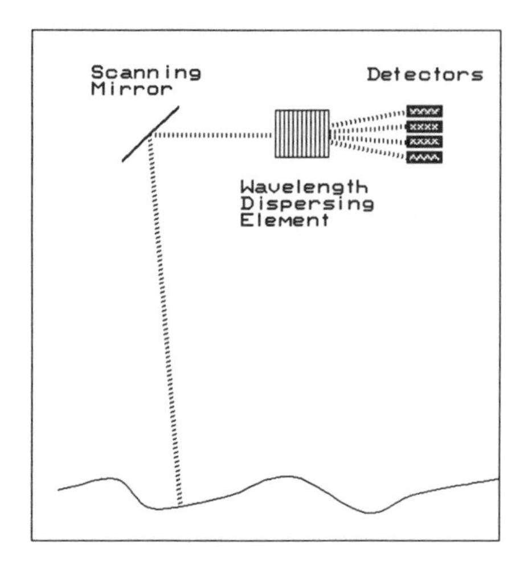

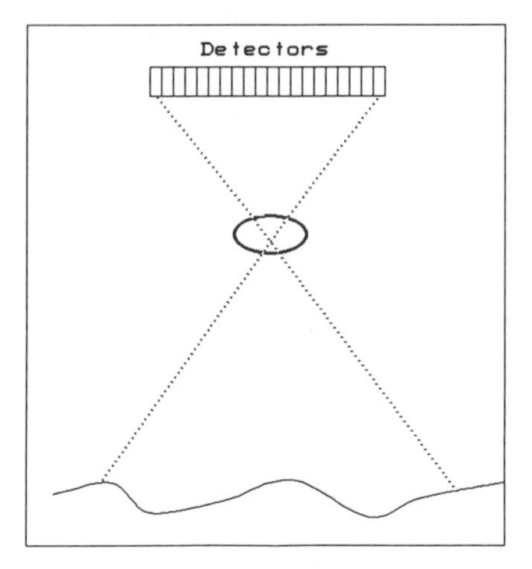

Figure 10.5 Whisk-broom versus Push-broom image scanners.

an active across-track scanning system. Also, since each detector views only a single pixel in a row of the array, the time available for collecting energy from the corresponding ground location is much greater. This greater dwell time has beneficial effects on the overall performance of the imaging system. To take data simultaneously in several spectral bands, we can enlarge the line of detecting elements to a rectangular or **area array**, and place optical dispersing components to break the spectrum into the appropriate locations in the system so that successive rows in the detector array correspond to different wavelength intervals.

Data from such systems is typically organized as band-sequential or band-interleaved raster datasets. Unfortunately, a common complication is the inclusion of supplemental data of various kinds into the data file, as well as dividing the geographic coverage into separate areas. For example, consider an instrument with four spectral bands, each band represented by an 8-bit byte, which has imaged an area into 1000 rows of 600 pixels each. A common organization of this data is to write four sequential raster files, one per band, with each file consisting of 1000 records of 600 bytes each. Another simple alternative is to write one band-interleaved file. However, the band-interleaved file could be arranged in at least two different ways. The first alternative is to use 1000 records of (600 pixels time four bands) or 2400 numeric values per record. Thus, each physical record of 2400 bytes corresponds to one logical record, since each single record is the entire set of observations for that row of pixels. The second way would be to record 4000 records of 600 pixels each, so that four physical records, one per band, correspond to the logical record.

Adding the ancillary data records complicates the matter further. As a rather extreme case, we could imagine the dataset above stored as 37 separate files: a single overall descriptor file that records information about the entire image (such as the date and time of acquisition, sun intensity and viewing geometry at that time, estimated cloud cover, and platform altitude), and then four sets of nine files, one for each geographic quadrant of the data. Each set of nine files starts with a quadrant header file that indicates the geodetic location of the quadrant's four corners, followed by two files for each of the four spectral bands: a band header file with information about the sensors for that band (including calibration data, for example), followed by the data itself. We have also seen instances in which an additional set of files is added to the set that contains arithmetically synthesized data of known characteristics (such as a continuous light-to-dark gradation across the width of the image), so that we may test our programs that attempt to read the data!

The tools of remote sensing are the sensors themselves, the data delivery system, and the techniques of image processing. A brief description of some of the best-known satellite sensor instruments appears in Table 10.1. Note that the spatial resolutions and scene sizes are nominal, based on the respective agency's normal data processing and distribution procedures.

Table 10.1 Characteristics of Some Remote Sensing Systems.

Sensor Name	Operational Dates	Spectral Bands	Spatial Resolution	Scene Width
Landsat MSS	1972-present	4	80 m	185 km
Landsat TM	1982-present	6	30/120 m	185 km
SPOT	1986-present	3	10/20 m	117 km
GOES VISSR	1975-present	6	6.9 km	hemisphere
NIMBUS AVHRR	1978-present	4,5	1.1 km	2700 km
Heat Capacity Mapping Mission	1978-1980	2	600 m	716 km
Seasat SAR	1978	1	25 m	100 km
Shuttle Imaging Radar	1981 & 1985	1	917-58 m	20-50 km

10.2.3 Multispectral Data

Implicit in a multi-band approach to remote sensing data acquisition and analysis is the concept of **spectral signature**. The basis of this approach is that different materials exhibit unique reflectance and emission properties as a function of wavelength across the electromagnetic spectrum. By examining data from different wavelength regions, in an appropriate combination, a great deal of additional information can be extracted from multispectral imagery, compared to the amount that might be extracted from the same area using panchromatic imagery acquired at the same time. More specifically, underlying the concept of a spectral signature is the hypothesis that individual materials (and thus, objects or features) may be identified based upon their unique spectral response (hence *signature*) in a set of discrete wavelength regions.

Digital image-processing techniques have been applied extensively to multispectral imagery as an additional aid to information extraction. Numerous operations such as contrast enhancement, spatial filtering, and systematic noise removal have been used to improve the visual appearance of imagery, thereby improving its overall quality for human or manual interpretation. Also, statistical pattern recognition techniques have been used to classify or label land-cover types based on the spectral signature concept, using methods as we discussed in section 8.1.1. These pattern recognition techniques employ a variety of statistical methods to assign each pixel within an image to a specific land-cover class.

Despite the elegance of the spectral signature concept, the approach has several significant weaknesses. Frequently, the spectral responses of features on the Earth's surface exhibit high spatial and temporal variability. Therefore, the spectral signature of an object determined at one place and time is often not the same at other locations or times. For example, wheat fields and rural and urban landscapes in northern Kansas may not exhibit the same spectral characteristics as wheat fields and rural and urban land-cover classes in southern Kansas. This may be due to differences in farming practices and crop strains, variations in sun angle due to latitude, interactions between soil distributions and rainfall patterns, and many other factors. Also, the limited spatial resolution (in terms of the effective instantaneous field of view) of current remote sensing systems is such that a given picture element, which is the smallest area for which we have a unique data value, may not correspond to an area of homogeneous land cover. For example, a 10-meter pixel, when placed inside a 1-hectare homogeneous field of corn, provides representative brightness information about the crop in the field. The same pixel, when the field of view includes a portion of a paved road, plus the edge of the field, plus a corner of the barn, is much more difficult to interpret. It belongs in no single class, and should not be expected to.

This type of **mixed pixel** situation, in which a unit resolution element of a remote sensing system includes a number of land-cover types, is common in many remote sensing applications. Furthermore, the mixed pixel problem causes a number of serious limitations to traditional computer-assisted image analysis techniques. Consequently, computer-assisted image analysis techniques have generally been found to be inferior to human analysis techniques, in terms of both speed and accuracy. On the other hand, considerable work has been accomplished in this area, and as discussed in the following sections, digital processing of remotely sensed data holds considerable promise for application to GIS technology.

The visible part of the electromagnetic spectrum has wavelengths from 400 nanometers (blue) to 700 nanometers (red). Wavelengths somewhat longer than 700 nanometers are considered **infrared**; near-infrared has wavelengths of 700 nanometers to approximately 2 micrometers; middle-infrared has wavelengths between 2 and 5 micrometers; and the far-infrared has wavelengths of roughly 8 to 15 micrometers. Notice that there are gaps in this description. This is because the atmosphere does not pass energy of all wavelengths, but is relatively opaque in several wavebands.

The near-infrared portion of the electromagnetic spectrum is very sensitive to properties of vegetation, due to the spectral reflectance characteristics of the principal plant pigments and water contained in the structure of a leaf. Remote sensing of vegetation is a common practice, either when an application requires such information as plant biomass or species types, or when vegetation is a covariate for other characteristics of the earth's surface. It is common to examine a derived spectral ratio when plant abundance is a key concern. A ratio of near-infrared reflectance to red reflectance has been shown to be a strong indicator of green-leaf biomass (Hardisky, Smart and Klemas, 1983).

The near-infrared and middle-infrared portions of the electromagnetic spectrum have been shown to be of particular value for geological applications (see, for example, Goetz, Rock and Rowan, 1983). Many minerals show characteristic reflectance spectra in this part of the spectrum, which in semi-arid or arid conditions may be sensed at altitude. By comparing remote observations of spectral reflectance to measurements made in the laboratory, it is possible to be able to distinguish a number of minerals from each other (Goetz et al., 1985). In more temperate climes, geologists employing remote sensing are increasingly using vegetation as cues in their search for mineral deposits.

The far-infrared portion of the spectrum may be used to determine the temperature of an object. A number of studies have shown the utility of remotely sensed measures of Earth surface temperature. Such data can be of value when mapping surface currents in water bodies (LeDrew and Franklin, 1985). Time-varying surface temperature, for example, provides indirect information about soil moisture (Vlcek and King, 1983).

The microwave portion of the spectrum is generally defined as wavelengths of 0.1 to 200 centimeters. Active systems are common at these wavelengths, typically operating by transmitting short bursts of microwave energy, and then receiving the reflected energy on the same antenna used for the transmitter. Energy will be reflected efficiently when the target is smooth and perpendicular to the incident beam. A rough surface will scatter the beam in many directions, and thus, the amount of energy returning to the detecting system will be low. The lowest energy return will take place when the surface is both smooth and oriented to reflect the microwave energy in a direction away from the receiver.

The spatial resolution of a microwave system depends on the antenna system. For a simple antenna design, a longer antenna provides greater spatial resolution. For the many circumstances in which very long antennas are not practical, a **synthetic aperture radar** (SAR) uses complex signal processing techniques to simulate a long antenna. Synthetic aperture radar systems flown on the U.S. space shuttle have achieved 30-meter spatial resolution. It is feasible, with existing technology, to realize even higher resolutions with these kinds of systems.

Although still largely experimental from an applications perspective, the day/night and nearly all-weather capabilities of synthetic aperture radar have generated much interest in many communities. These systems have been used by mineral exploration companies working in the tropics, where cloud cover seriously hampers the use of aerial photography. The Brazilian government has used an airborne SAR system to image millions of square kilometers of the Amazon basin. These images form the basis for a series of 1:250,000 scale maps of the area.

10.3 Digital Processing of Remotely Sensed Data

Human beings are very talented image analysts. Why should we try to improve on their abilities? There are a number of reasons. Computer-assisted processing can sometimes distinguish between more subtle differences in a target than a human analyst can. Whereas a human interpreter may be able to perceive some eight to sixteen gray levels when analyzing a continuous-tone black-and-white image, many more levels can sometimes be distinguished by employing the capabilities of machine-assisted analysis. For example, if an image is recorded with 8-bit quantization, it will have 256 unique gray levels, and may have more useful information than can be extracted visually.

Human interpretations may not always be consistent or repeatable. Image analysis is subjective, and as the complexity of the analysis process increases, the likelihood of an analyst being able to repeat a given interpretation with a high degree of fidelity decreases. Computer-assisted analyses are almost always repeatable. This is true even when the results may be questionable, or when the system is unable to spectrally separate the classes of interest with a high degree of accuracy based upon the available multispectral data. In addition, machines are much more capable of keeping track of the tremendous amounts of detailed spectral information found in digital imagery than are human interpreters. For example, a single image from the Landsat Thematic Mapper system contains over 200 megabytes of data, separated into seven spectral bands, and covering an approximately square area on the Earth's surface 185 kilometers on a side. If an analyst is tracking the spectral characteristics of the vegetation in a given scene through time in order to identify crops, estimate biomass, or to predict yields, computer-assisted analysis may prove a much more useful tool for accomplishing this type of analysis.

Finally, when imagery is acquired in a digital format from a remote sensing system, it is easier to apply certain types of corrections to the data. These types of corrections can be radiometric or geometric in nature. For example, it is relatively easy to change the display of Landsat or SPOT image data from one standard projection to another (see section 6.6). There are also procedures to correct substantially for atmospheric haze in an image, or to adjust for changing angles of solar illumination for imagery of a given area acquired at different dates. Again, this is not to imply that computer-assisted analysis of remotely sensed data is superior to human analysis: it is not. It may, however, be better suited to different kinds of problems.

The tasks of both human image interpretation and computer-assisted image analysis are essentially the same: the detection, identification (or classification), and measurement of image features, and problem solving based on the identified objects. However, very little contextual information is normally incorporated into the decision structure of current machine-assisted image analysis algorithms. In particular, these procedures generally attempt to process digital imagery solely on the basis of tone or color and primitive measures of texture. This varies from the human approach, which incorporates the full range of image elements (discussed in section 6.8).

Early computer-assisted analysis research emphasized statistical pattern recognition approaches for classifying multispectral data. The mathematical elegance of these techniques is highly appealing and often led to overly optimistic expectations for their general utility and applicability to all types of remotely sensed data. In many cases, a variety of evidence from both the image and auxiliary sources must be assimilated in order to derive an accurate interpretation of an image. It is now widely recognized that pattern recognition techniques alone are often inadequate; many situations require contextual information or prior knowledge about the scene in question, both of which are basic elements used in human image analysis. This realization has led to a great deal of interest in techniques developed by researchers in the artificial intelligence and computer vision fields (see section 11.4). These techniques include image understanding, symbolic reasoning, and knowledge-based image interpretation approaches. Although not unrelated, these fields each contribute distinct features that are relevant to computer-assisted analysis of remotely sensed imagery.

In the final analysis, when an image analyst is going to perform an interpretation task, there are a number of important image-processing considerations that must be taken into account. The analyst must review all of the available options as they relate to a given application or problem. The analyst must then gain as full an understanding as practical of the problem being addressed, and then identify those objectives and methods that are to be used to extract the required information. If the use of remotely sensed information is appropriate, the analysts may have alternatives that need to be considered. For example, the analyst might wish to purchase existing coverage of a given area. This data might be in analog or digital form and might be obtained from a map and imagery library at a university, or purchased from either a commercial firm or a government agency. If we assume that image processing is to proceed in digital fashion, a number of steps must be taken. These steps are listed in Figure 10.6

Here an example will serve to illustrate the process. The example is that of a land-cover-change detection problem. Initially, the user must determine the type of questions that need to be answered. The user must then select and acquire the data types and platforms from which the information will be extracted. In our test case, this may involve the purchase of copies of historical aerial photography from a local government agency, plus ordering new satellite coverage from a commercial firm. The user must then decide upon the bulk and specialized preprocessing steps required for the given task. This choice can range from a set of radiometric and geometric correction algorithms to specialized band ratios, edge-detection, and masking procedures. Continuing with our land-cover-change example, a contract for scan-digitizing the historical photography might be let, to convert this non-digital data source

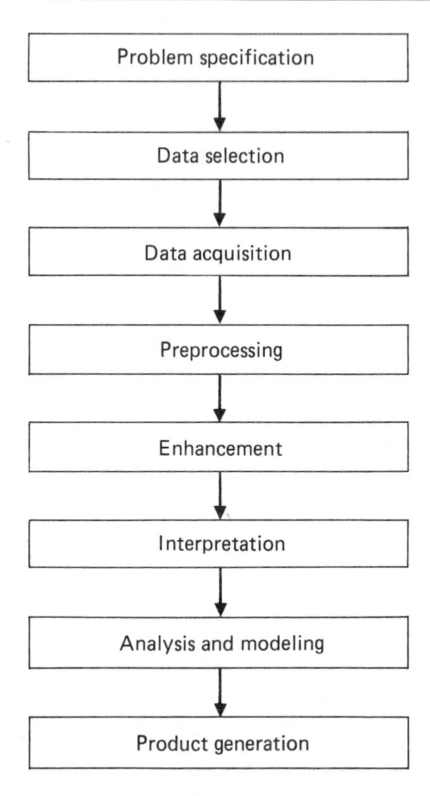

Figure 10.6 Analysis processing steps.

Continuing with our land-cover-change example, a contract for scan-digitizing the historical photography might be let, to convert this non-digital data source to a form more easily handled by the project staff. Alternatively, the aerial photographs may require manual interpretation and the use of an optical transfer scope to transfer the interpretation to a map base before digitizing. At this point, a digital registration process brings the two dates of imagery into a common projection and scale, and then radiometric enhancements may be applied to each of the images.

Once the registration and enhancement processes are completed, applications modeling and data analysis are quite often an iterative process, as preliminary results are tested and rechecked. For our example, various different classification algorithms may be applied to each of the images to try to extract the key land-cover classes for analysis. Once these results have been computed, they may be displayed in a variety of forms appropriate to the demands of a given decision. Once a decision has been made and a course of action has been taken, further local, regional, and global changes begin to manifest themselves, and the cycle of change, detection, decision and response begins again. One important point that we reemphasize here is that remote sensing provides an important tool for maintaining the currency of important data layers over time.

One of the key functions in applications of remotely sensed data is that of classification. The data obtained from most remote sensing imaging systems is continuous, in terms of brightness in one or more defined wavelength bands. On the other hand, much of the data processed in a geographic information system is organized as sets of discrete categories, such as categories of land use, or species of vegetation, or discrete administrative districts in a region. We have introduced the two different classification strategies in section 8.1.1. These techniques for converting the continuous remotely sensed data into discrete categories are extremely common as functions for interfacing image processing and geographic information systems, as we will discuss shortly.

In addition to classification as a tool for extracting similarities in datasets and converting between different systems of description, the classification function provides a means of data compression. For example, since there are limits to our ability to discriminate between colors on any practical image display system, we can often reduce the total size of a dataset without reducing the apparent information content of the scene. We might take a three-band red-green-blue image display and use an unsupervised classification algorithm to create a new image with 256 unique categories. We may then display each category in the classified image in a color that is representative of the category's average original color. Based on our experience, one often can't see a significant difference between the original image, which requires 24 bits per pixel, and that which has been compressed through classification in this way, which requires only 8 bits per pixel.

10.4 Interfacing Remote Sensing and Geographic Information Systems

As we have mentioned, interconnecting remote sensing systems to geographic information systems is valuable in many different applications. We will now discuss a number of common techniques for moving data between these two related kinds of spatial data-processing systems.

Remotely sensed data is almost always processed and stored in raster data structures. When working simultaneously with an image-processing system and a *raster* geographic information system, it is usually easy to move data between the two. Typically, a single theme of information is extracted from the (often multispectral) remotely sensed data. In work concerned with vegetation, a ratio of brightness values in different bands (typically contrasting infrared and red spectral channels) may provide information about the abundance of green vegetation. Other ratios may highlight different soil and rock compositions. Temperature differences between day and night at a particular location may indicate the moisture content in soil. The remote sensing discipline has developed a number of different techniques for transforming general-purpose datasets into thematic information for many related fields.

In other instances, discrete categories of surface cover may be distinguished through a classification algorithm, discussed in section 8.1.1. Many different decision rules may be developed to isolate or characterize components of the Earth's surface. Often, a sequence of discriminating steps is required to provide an acceptably accurate or precise result. For example, a simple binary decision rule may be able to exclude water bodies from consideration. A second binary decision may then exclude snow-covered ground and clouds. Finally, an unsupervised classification can then permit us to focus on cultivated vegetation. This example is oversimplified, but illustrates a general principle that is parallel to the use of interpretive keys in photointerpretation (see section 6.8), where we successively prune away the uninteresting alternatives.

Once the remotely sensed data has been converted to a desired data type, transferring this data to a raster GIS is relatively simple. Header and trailer records or files may need to be modified during the conversion process, but converting between different raster data structures, as we discussed in section 6.1.1, is relatively easy. In our experience, most operational image processing and raster geographic information systems provide mechanisms to read and write 8-bit-per-pixel raster arrays.

More work is involved when transferring raster data derived from remote sensing systems to a vector-based GIS. In the following, we present only the outline of the process. One possible sequence, based on derived continuous data such as vegetation abundance, involves extracting the contours of abundance (often called isolines) on the image-processing system, and converting these raster representations of contour lines to vectors (perhaps using techniques like those mentioned in section 6.1.1). The vectors may then be passed to the GIS, along with labels to indicate values associated with the contour lines. When working with information which has been classified into discrete categories, such as land use or rock type, appropriate transformations are made to convert the continuous multivariate brightness data into discrete categories. Then, to isolate the implicit homogeneous polygons in the derived image, the pixels that form the boundaries of the areas are detected. This is exactly like the skeletonizing algorithms mentioned in section 6.1.1. These boundary pixels may be used to develop the vectors surrounding the areas, and attribute values and class names are assigned to the bounded areas.

It is important to understand that the conversion processes are limited by the underlying data, in terms of precision and accuracy. A 512-by-512 array of 24-bit pixels, displayed on a high-quality color monitor, may be very satisfying to the analyst. However, the vectors that are developed from this discrete array may look poor on a high-quality plotter, since the plotter has many more addressable positions (and thus, higher resolution) than the image display screen. Often, the analyst will smooth the plotted lines on the final product to make a more pleasing presentation. If a user smooths the boundaries between classes of spatial objects found in the image to make the vector plots appear more realistic, he or she must *please* remember that the underlying data may have been irreversibly changed.

10.5 Synergism

We will use a small number of applications examples to illustrate some of the synergisms that lie at the margin between remote sensing and geographic information systems. In a number of areas, there are important benefits to be gained when we bring both the detection and monitoring abilities of remote sensing as well as the philosophical approach and analytic capabilities of a GIS to bear on a problem.

A key area in the joint applications of remote sensing technology and geographic information systems is to identify change. Whether this change is of interest for its own sake, or because the change causes us to act (for example, to update a map), remote sensing provides an excellent suite of tools for detecting change. At the same time, a GIS is perhaps the best analytic tool for quantifying the process of change.

Typically, the basis of change detection is the comparison of remotely sensed data and map data, or the comparison of remotely sensed data taken at two or more different times. When working with multi-temporal data sets, geometric registration is a fundamental concern. If we are unable to register the two images with high precision, the errors in registration will make it appear as if there has been some change in the landscape. Consider a simple problem, such as a road running through a grassland. If this area is imaged on two dates, and the imagery registered to each other imperfectly, some pixels that were originally grass will appear to be road in the later image, and some pixels originally road will appear to have changed to grass. Without critically examining the quality of the registration process in terms of accuracy and precision, we could be fooled into believing that the roadway was moved during the interval between the images. This is equally a concern when the images are each rectified to a common base map, rather than registered to each other.

Once the images are registered, there are two alternative methods to quantify change. The conceptually simple approach is to compare the pixel values in each of the images. However, while this is simple to calculate, it is exceedingly difficult to interpret. Dry sand at the edge of a reservoir might be submerged at a later time; the pixel thus gets generally darker. A field of corn goes from peak greenness in summer to light debris after harvest; it gets much lighter during this interval. This of course ignores the details of complex spectral changes. If the images can be registered with very high precision, and if the dates of the images represent identical moments in the seasonal cycle of vegetation growth and illumination, and if the viewing geometries of the two images can be taken into account, and if a host of other unlikely elements are precisely the same, then it is reasonable to compare the pixel multispectral brightness values between the two dates.

The common alternative is called **symbolic** change detection. The analyst first decides on a set of thematic categories that are important to distinguish for the application. For example, we may be interested in change at the margin of a city, to keep track of expanding development into rural areas. In this case, the categories of interest refer to land use and land cover, and the objective of the exercise might be to examine transitions from undeveloped categories (such as grasslands and forest cover) to developed categories (such as golf courses, housing tracts, and shopping centers). As above, the two images must be registered to each other, and again, it is important to understand the residual errors in the registration process. Next, each of the two images is classified independently, according to the same set of categories. In this way, control for the effects of sun angle, changing sensor sensitivity, and seasonal patterns and change in surface features is developed separately in each image. The final comparison is made on the classified data, from which it is easy to generate such output products as a map of those areas that show any change at all, statistical tables describing the change in areal extent of the different categories, and a matrix representation showing the probabilities of change from any category to any other.

It is a small logical step to use a map in place of one or both of the images in the examples discussed above. Again, the problem of registration between the data from the two (or more) dates is a key operational concern. In essence, the classification process described above generates a map-like representation of the data, in that we have greatly reduced the dimensionality of the original data into a set of categories specific to the problem. In some cases, the map itself may be scanned to create a raster dataset, and thus, enters the processing flow much like any other image.

In section 8.8 we discussed a study that centered on the universal soil-loss equation. This is another application in which remotely sensed data forms one input to a geographic information system, in this case with a number of others (such as soils, slope, and rainfall).

Another family of applications in this general area is termed **map-guided image interpretation**. This refers to the use of maps as an aid to interpreting imagery. While this is common practice for human interpreters, there are methods in which map-based data is of great value in an automated image analysis process. Tailor et al. (1986) describe three separate methods for integrating maps in this way: stratification, classifier modification, and postclassifier sorting.

Stratification involves incorporating the map data before classifying the imagery. The region is subdivided into smaller areas, based on the map, and each of these areas is then processed separately. As an example, in rough terrain there may be areas that are in shadow. Based on the elevation data in the map, and on information about the sensor's viewing geometry and the position of the sun, we may be able to predict where the shadows should fall in the image. Thus, we can process the shadowed areas separately from those areas that are brightly lit. Without such a procedure, the same vegetation may be placed in different categories, depending on whether the site was in shadow during the time of image acquisition. From the view of a very different application, sites in a township may have more or less value for a project than sites outside the township boundaries. By using the map to determine the location of the key boundaries, we may be able to remove the inappropriate areas before beginning the bulk of the image-processing computations, and thus decrease the costs of the analysis.

Classifier modification involves including information from the map during the classification process. In effect, the classification algorithm acts on both the remotely sensed data and relevant information from the map. In other words, the map data forms another logical data channel for input to the classification. Elevation and slope, for example, might be as relevant to a classification for a particular purpose as a spectral channel. Vegetation types stratify into zones based on elevation in an area, and as such, elevation may be a valuable variable to include in the classification step. We caution, however, that mixing nominal categories of information (such as land use, or ownership, or presence within a given administrative district) with continuous ratio variables in a classifier that assumes the latter may produce misleading results. This is a problem parallel to that of interpolation with categorical data, mentioned in section 6.6.4. When working with categorical data for assistance in classification, it is common to extract the image data from each class, put it into separate files, and then classify the pixels from each class in separate computer runs.

The third method for incorporating map data to aid in image interpretation is called **postclassifier sorting**. This involves modifying the results of the spectral classification by relying on the map (or other ancillary data). A common application of postclassifier sorting is to determine the identity of the spectral class "miscellaneous." Often, there will be a collection of pixels whose identify is unknown, possibly because of an insufficient number of training fields (in a supervised classification). In cases like this, the map can provide some guidance on the identity of these pixels.

In the final analysis, users of remote sensing and GIS technologies have moved from analyses that were aimed at simple feature identifications to analyses used in complex problem solving. Examples of this trend range from the impacts of changing climates on vegetation, to the use of reverse trajectory hydrologic modeling, in conjunction with well log data and historical photography, to locate sources of groundwater contamination in urban aquifers. These examples can also be used to illustrate the changing nature of the spatial scale of analyses as well. From the local analysis of urban aquifer contamination we have, with the advent of the acquisition of satellite data, moved to analyses that are global in scope. These and similar images are being processed to understand processes that affect the long-term habitability of the Earth.

To accomplish this will require that we collect data over long time periods, probably on the order of decades, and that we combine these remotely sensed data with data collected on the ground, in the atmosphere, and even from subsurface strata (e.g., soil moisture or depth to groundwater) if we are to address the issues in a full and effective fashion. Such processing implies an information systems approach wherein a key element is a geographic information system.

Chapter 11

Practical Matters

Given our discussions in the previous chapters, it should be clear that the design of a spatial data-processing system is a large and complex task (Smith et al., 1987a). Even though few of us will ever design such a system, an appreciation of some of the key design issues in Chapters 4 and 7 can be valuable for a system user as well. At the same time, procuring a system can be a significant challenge. Installing a geographic information system has a number of implications for any organization. The following discussion starts with material taken from a meeting several years ago, where we were invited to help a manager decide on the purchase and installation of a geographic information system. We then describe several other practical matters that affect the operation of geographic information system for solving real-world problems.

As you will see, sometimes practical considerations can lead to some unexpected results.

11.1 A Case Study

The U.S. Marine Corps Combat Center at Twenty-nine Palms, California, is a large area with a variety of users and uses (Star et al., 1984). The base covers some 600,000 acres in the California desert. One of the most important uses of the base is for Marine Corps training exercises, for which soldiers are brought to the base for short periods of time. Some time ago, we were invited as part of a group of specialists to a conference with the base natural resources manager. The gentleman in this position at the time described one of his principal duties as meeting with training officers, to help decide where to place particular training exercises. Assigning these activities to specific areas on the base required him to take into account the needs of the exercises as well as the effects on the base. This can be considered a specific example of land-use management under frequent and changing demands.

The Combat Center natural resources manager was under increasing pressure both to manage land use on the base more effectively, and to take less time to reach decisions. He was familiar with several high-technology aids for his work, since the base has had good relations with the U.S. Environmental Protection Agency (EPA) laboratory in Las Vegas, Nevada. The EPA laboratory was very familiar with aerial photography and other kinds of remote sensing tools, as well as geographic information systems. The natural resources manager had a small budget to allocate, and he wanted to consider whether he could purchase a geographic information system. He hoped that a system might be available that could process and display remotely sensed data as well. After reading brochures from several companies and discussing options with manufacturer's representatives, he identified a system that was within his budget.

Our meeting took place over two days. On the first day, we attempted to determine the demands that would be placed on a spatial data-processing system for the base, in terms of data volumes, data types, and any requirements on **throughput** (the amount of data processing that must be completed in a given time). However, the personnel who would become the users of the system were not thinking in detail about some aspects of these problems, and indeed, did not speak in the same terms we used. At the same time, we, as the outsiders, did not understand the constraints on the staff or on the system itself. We would ask, "How fast must the system be able to solve significant problems?" They would answer, "Very rapidly." We would ask, "How much data do you have?" and they would answer, "A great deal of data." Things were not going well, and both groups were very frustrated. We, as the outside advisors, did not feel our time was being used efficiently. They, as the future users of the system, did not feel that we were working towards a useful input to their decision-making process.

On the second day, we took a new approach. We asked the base staff to describe a problem they were given, for which the resources at their disposal were not sufficient to solve the problem in the time that was available. The example they described involved a training exercise with tracked vehicles. The natural resources manager would be told that in 24 hours a group would arrive on the base and would begin learning to drive tanks through the desert. The environmental staff would need to provide information to the training officers on those geographic areas where the vehicles could not go, because of the presence of habitats of rare and endangered species, or because of sites of archaeological significance. Based on this information and the suitability of different kinds of terrain for the vehicles to be used in the exercise, a decision would have to made on where to conduct the training exercise. This specific operating scenario became the key to our understanding the problem, as well as to our making some specific recommendations.

As a starting point, we asked about the staff to develop a list of the different kinds of data they maintain. For natural resources management, they were able to describe some 30 identifiable thematic data categories, including elevation, soils, land use, location of habitat for a number of different plant and animal species, and locations of recognized archaeological sites. The data for many of these layers were normally recorded on flat maps. A serious first problem would be turning this large number of maps, which includes portions of 42 different U.S. Geological Survey 7.5-Minute quadrangle sheets, into a consistent digital database of verified accuracy.

Second, we tried to assess the size of the digital database that might be created by entering the data into a digital geographic information system. As a first step, we assumed a simple raster data model, with each theme stored in a separate raster array. Such a data structure was used in the commercial system under consideration. We asked base staff about the required precision of this database, to establish the underlying raster cell size. Considering the tank exercise again, the base staff said they were comfortable if they could indicate a 50-meter impact zone around key sites where tank traffic would be forbidden. Based on this figure, we estimated a raster cell size of 12.5 meters, so that 50 meters, plus or minus the width of one cell (or 12.5 meters), still provided adequate protection for archaeological sites and rare species habitats. From this desired cell size, and the total area of the base, we estimated that 16 million pixels would be needed per data theme, and thus (making some assumptions about the dynamic range of each data layer), a total of roughly 200 megabytes of digital data for the 30 data layers.

This thought process was sufficient to make a recommendation. The system under consideration for purchase stored data on 8-inch standard floppy disks, holding approximately 160 kilobytes of data each. From our rough calculation of the database size, we estimated that the database would require some 1200 floppy disks for storage -- if stored in standard containers, this is a single shelf 30 meters long. This is probably not a reasonable alternative. Note that this calculation is certainly conservative, since it ignores one or two generations of data backup, which would double or triple the required number of disks. Furthermore, the process of generating maps of perhaps 10% of the area of the base, considering a dozen data themes for generating maps of exclusion for the training exercise, could not be completed in the allotted 24 hours. In addition, the system did not have reasonable capabilities for the display and manipulation of remote sensing datasets. Our final recommendation was to do nothing: the system that was economically realistic could not meet the specific requirements of the task.

There are a number of interesting studies that examine the utility of automated spatial data processing systems in the recent literature. As a counter-example to the one discussed above, Hansen (1987) discusses a situation in which installing a GIS was clearly the correct choice. Goodchild and Rizzo (1987) present a more formal analysis of estimating the performance of a GIS. In their model, they use a regression analysis to develop equations for predicting the costs of several stages in processing (such as labor costs for manual preprocessing, costs of converting the scanned maps to a vector dataset by contracting with a data-processing service agency, and subsequent manual error corrections) as a function of the complexity of the map manuscript.

11.2 Database Volume

Estimating the volume of data in a geographic information system is often an important consideration when designing or specifying a system, and is particularly easy when considering simple raster structures for data storage. Light (1983) estimated the amount of storage that would be required to input the information in topographic maps in a digital raster database. We find his calculations instructive, as an outline of the method. The archetype map for this study was a U.S. Geological Survey 7.5-minute topographic quadrangle, at 1:24000 scale. Roughly 54,000 such maps cover the United States, excluding Alaska.

Light assumed that the minimum mapping unit (effectively, the smallest feature depicted) on these maps is 0.25 millimeters, and then decided that approximately 12 raster cells should cover features of this total area. Note that this corresponds well with our suggestions in section 4.1.1. For a map at a 1:24000 scale, this works out to a cell size on the ground of 1.7 meters on a side. The 54,000 maps cover a total area of 7.84 x 10^{12} m^2, which works out to approximately 3 x 10^{12} cells. He determined that the raster data would require seven different themes, for a total of 34 bits per cell:

Data Theme	**bits/cell**
Panchromatic Image	8
Vegetation	2
Hydrography	2
Hypsography	2
Road Class	2
Cultural Features	2
Elevation	16
Total	**34 bits/cell**

Thus, multiplying the number of cells by the bits required to store the data for each cell, we arrive at a total of 9 x 10^{13} bits. The author considers this a liberal estimate, since we are computing the simplest raster data structure. Thinking in terms of placing this amount of data on a computer storage medium, Light calculates that it would require 330,000 9-track 1600 BPI magnetic tapes, or 3100 optical disks. He further points out that the lifetime of the magnetic tapes, on the order of three to five years, prevents us from using such tapes in a practical sense, because of the costs of maintaining such a large tape library.

One could certainly argue that this calculation is naive, since it ignores possible efficiencies such as vectors for some features and any of the raster compression techniques mentioned in Chapter 4. However, the sequence of steps provides a guideline for similar calculations.

We would like to reinforce the above comment about the lifetime of magnetic tape. Magnetic tapes do not represent a static archival storage medium. Data stored on ordinary magnetic tape is volatile; tapes left alone in a rack, even under optimum environmental conditions, deteriorate over time. The data stored on the tapes typically last only 3 to 6 years. To ensure the long-term integrity of a tape data archive, it is important to clean the tapes regularly, and ultimately to transfer the data to new tapes. This data management function is crucial to the long-term viability of any digital dataset, a fact that is well-known to the staff of any large information center. Unfortunately, the impermanent nature of magnetic tape storage is unknown to many new users, and the need for continual maintenance is often not considered when implementing new spatial data-processing systems.

11.3 Specifications

Questions about time enter discussions of spatial data processing at several levels. From a simple point of view, we often have a temporal performance specification: a client needs an answer to one of a family of questions in a finite period of time. A dispatcher of emergency vehicles may have only a few minutes to be able to advise fire and police services that there are known hazardous materials stored at the scene of a fire. A real estate developer or investor may need to know the relative suitabilities of the prospective sites before the end of the tax year, in order to make financial decisions. These specification are end-to-end requirements, in that they include understanding the problem, any needed data acquisition, preprocessing, and so forth.

From another point of view, once some data has been acquired it is by definition out of date. How often is database update required for a particular kind of data? For a particular application? When there are explicit data accuracy specifications for the different data layers, the practical answer is based on these standards: when a dataset no longer meets the stated accuracy standard, it is time to update the data. For example, a dataset may originally have been created with an accuracy criterion based upon a specified comparison with field observations. Based on a random set of comparisons between field observations and the stored data, there must be agreement between the electronic data and the field observations at a specific statistical confidence level. If the data quality was originally established with such a mechanism, the existing dataset could be tested periodically with new field observations to determine whether the data meets the original specifications, and to determine when an update cycle is in order. This reinforces the importance of standards for accuracy of spatial data.

Establishing the precision of a dataset is a difficult practical matter. It is simple to determine the numerical precision of stored data values: elevation stored to the nearest *centimeter*, population density recorded to the nearest *person per hectare*. However, it is difficult to determine the level of precision of the underlying *information*. Consider a simple array of bathymetry (depths below the water surface), originally based on a near-shore sonar survey (we temporarily neglect errors in horizontal positioning, which may be very substantial). Along the track of the acoustic survey, we have computer-stored values of depth, based on the acoustic measurements plus assumptions about the velocity of sound in water during the survey. These stored values may be recorded to the nearest tenth of a meter, a choice of precision due more to the word length of the computer than to either calibrations of the system or our faith in the equipment.

Interpolations are used to take a series of measurement runs with the positioning equipment, and to develop a raster array of depth values. The interpolation procedure is certainly precise, in the sense that the calculations are performed to a great many digits of numerical significance. But after all of the assumptions about physics (the velocity of sound traveling through water), engineering (the design and practical operation of the acoustic profiling system), and physiography (which hopefully was taken into account in the design of the interpolation function), does anyone really know about the true precision of the resulting dataset? This is a difficult problem.

Errors are rarely considered explicitly when using a geographic information system. The input data may have some information about accuracy and precision. For example, confidence intervals may have been derived from the responses in a demographic survey. But when the final map of population and education trends over the city is plotted, where is this confidence information? The same comments apply to any manually drafted map: the cartographer may have made an expert decision to place a feature on the map at the "most likely position." Should we then ask the cartographer to use a fuzzy line when there is sufficient doubt? When the digital elevation model is developed from a set of surveyed elevation point plus a manually digitized map, the root-mean-square error of the derived data points may be estimated. Based on this internal consistency measure, how much do we *trust* the individual data values? What measures have been used to guard against errors of various kinds? How does error in the elevation model propagate into a derived data layer, such as local slope? And how does the error in the original elevation data affect the client's confidence in an expensive decision to move many cubic meters of earth for the foundation for a construction project? These are areas of active research in the community.

From a different point of view, we acknowledge that there are times when we are simply not using all the tools at our disposal. This may be due to a lack of time, or a lack of staff; unfortunately, however, it is sometimes due to a lack of creativity and foresight. For example, in some instances we may have knowledge of the surveyed locations, which were used to calibrate and verify the accuracy an elevation dataset. Based on this information, we could derive a synthetic confidence data layer in the following manner. We would mark locations as "high confidence" when they are near a surveyed point, mark locations as "low confidence" when far, and mark locations "reasonable" when their estimated elevation is consistent with our knowledge of the local land forms and regional physiography. This ancillary data layer can be used in two useful but distinct ways. First, it can help document our knowledge of the elevation dataset, and second, we may be able to use this layer to infer confidence in derived products later on in the information flow. To complete this scenario, in a raster-based system, the confidence data would be recorded in a separate data layer, while in a vector-based system, the confidence metric would be an additional attribute of any spatial object.

Continuing in this area, for a statistical classification of ground-cover classes based on multispectral scanner data, we often have some limited internal consistency measures. In many software systems, we can calculate both the data layer containing the class assignments for each pixel or polygon, and a measure of class separability for each pixel. If the system has calculated a 95% probability that the raster cell is *urban*, and a 5% probability that it is not, we can feel reasonably comfortable with the labeling of the cell as urban. If the probability is only 51% that it is urban, the system will still label the cell as urban -- but we are rather less comfortable with this result. There are two usual responses to the latter situation. Some users just ignore this information, assuming (hopefully with justification) that the most likely class is of use, even with relatively little confidence in the result. In some applications, however, it may be appropriate to develop a new ground-cover class: *we are not certain*. How soon will we have the wisdom to incorporate this information about confidence in the intermediate steps into specific statements about confidence in the overall decision process?

11.4 Project Management

Managing a spatial analysis project can be a complex undertaking. There are a range of problems to consider, including operating policies, long-range staff and facility development, documentation, and training. When a project includes introducing a new way of doing business, such as adding a digital geographic information system, there is the additional problem of integrating the system into the existing institutional framework to get the job done.

The use of a geographic information system for satisfying a given need is called an application. Most GIS applications involve sophisticated geographical analysis, such as resource allocation or land-use planning and management. As such, most of these applications involve, to one degree or another, mapping, monitoring, and modeling activities.

The stages in any project (based in part on Hewitt and Koglin, 1987) include:

Identify the client's needs:

The client may be another division within your organization or may be an external agency. In any event, the time spent in specifying the overall objectives is extremely valuable. Objectives should be specified with sufficient precision in terms of deliverable products, as well as constraints on time (such as performance milestones and due dates for intermediate products), budget, and staff, so that they may be converted unambiguously into a project plan.

Identify the underlying analytic model:

For the analysis task defined above, identify the analytic means of satisfying the objectives. This may range from a relatively straightforward count of occurrences of target species extrapolated from a field observation program, to a client's proprietary model of groundwater percolation and flow.

Identify the data requirements:

It may be possible to find some of the required data; it may also be necessary to develop other data resources from scratch. The decision process should include information about both spatial and non-spatial characteristics, as we have discussed them throughout this text.

Information about desired data accuracy and precision, from a quality assurance point of view, is critical.

Identify the processing system requirements:

Are needed algorithms not available on the target GIS? Are datasets not in suitable digital data formats? Is there an appropriate amount of data storage for the processing jobs as well as the project archives? Will hardcopy devices be required that are not on site?

Identify the staffing needs of the project:

Will there be requirements for staff members who are not available (i.e., applications programmers, additional system programmers, environmental biochemists)? Will locating such staff, either as consultants or otherwise, hinder our ability to finish the project on time? How will these problems affect the tentative budget?

A classic book on developing software systems should be in every library: *The Mythical Man-Month* (Brooks, 1975). While it has direct relevance on the problem of developing new software for a project, this brief volume has valuable general guidelines that are relevant to almost any technological undertaking. One of the key points made in this book is that the length of time required to finish a task is not a linear function of the number of workers and the amount of effort expended by each. When many people are working together, there is considerable time spent coordinating the staff. Thus, if a single person can finish a problem in one year, it is unlikely that twelve people could complete the problem in one month, and *extremely unlikely* that 365 people can complete the problem in one day.

Chapter 12

Applications

The following sections describe a number of applications of geographic information systems. The examples we present here are intended to illustrate the wide range of practical applications of this approach, using currently available technology. Among these are representatives of both raster and vector database structures, both costly and inexpensive systems, and both simple and complex analytical models.

12.1 Master Planning

Master planning applications often involve a broad range of data types and analyses, to develop a basis for the development of a property. The following description of a master planning application was provided by Intergraph Corp. of Huntsville, Alabama; the application was run on Intergraph hardware and software.

This example is based on a master planning problem: site selection, site development and facility design for a new medical facility for a U.S. Air Force base in San Antonio, Texas. The upper panel in Plate 2 shows the general characteristics of the Air Force's spatial database for the site. Five separate vector data layers, and their associated attributes, have been collected. These layers include a general base plan (legal boundaries, streets, buildings, etc.), the water supply system, the gas delivery network, the communications network, and the electrical network. Note that the database is not contiguous, but is partitioned into rectangular units to improve performance and maximize access by multiple users, as we discussed in section 7.4.

The master planner begins the project by identifying areas which either support or prevent the development of the planned medical facility.

Exclusion zones may be based on noise from airport operations, areas affected by aircraft landing patterns, and areas that would interfere with airport radar. These exclusion zones are created by site-specific models, and are modified by the expected dimensions of the proposed hospital building. For example, areas which are excluded from consideration due to noise are derived from the expected distribution of noise sources, the characteristics of the noise, and models of sound propagation. Additional exclusion zones are developed for the project from knowledge of flood hazards, based on local topography. An example of these exclusion zones, overlaid on a portion of the infrastructure map, is presented in the center panel of Plate 2.

After determining those portions of the base which must be excluded from consideration for the hospital, areas may be identified that might be suitable. Proximity to primary roads, access to utility trunk lines, relatively gentle terrain are factors that can support cost-effective and environmentally-sound site development. An overlay analysis then combines the exclusion and inclusion zones. The master planner compares the results of this site suitability analysis with information about current land use regulations and total area requirements for the new facilities to determine the best site for the medical building.

Site specifications are then provided to the site designers and architects. Networked computers and multi-user access allow site development and facility design to proceed simultaneously, and data sharing among these groups permits the users to verify design compatibilities.

The site designers create a digital terrain model of the existing site, based on contour data. From these data, the system's database and computer-aided design functions permit the design of the entrance road, bridge, parking area, and the hospital building itself. A composite 3-dimensional wireframe model of the site topography, updated with the proposed facilities, is shown in the lower panel of Plate 2. Detailed earth-moving calculations which are based on this 3-dimensional model are used to estimate site preparation costs. Additionally, the system may be used to generate construction material schedules and architectural renderings based on the hospital's detailed design database.

If the site database is sufficiently rich, as in this case it is, the system can even provide information about the availability of base housing for the associated increase in hospital staff. A simple database report function can identify the number of vacancies in medical barracks, and if a shortfall is found, the master planner can merge recent imagery of the base with the existing map base to find open space suitable for temporary medical barracks.

This is, of course, not a simple application. It illustrates what is possible when a large database, an extremely wide variety of analytic operations and software functions, and personnel with skills in many areas are brought together. We include it to show what is possible, and to contrast with some of the simple examples present in earlier chapters.

12.2 Proposed Dam Site

This example is a fictional lake siting study. It emphasizes the use and value of three-dimensional display of environmental data. The images, provided by ERDAS, Inc., of Atlanta, Georgia, were created on a raster geographic information system. The essential dataset is a raster of elevation data. In the upper panel of Plate 3 is a color-coded shaded relief representation of this elevation data. Contour lines in this presentation show the lines of constant elevation, and colors have been used to highlight the proposed inundated area (in blue) and the surrounding area that might be suitable for recreation (in green). In addition, the image has been modified to provide a three-dimensional perspective, as if the light source for this view is in the upper right of the image.

A more dramatic presentation of the terrain data is found in the center panel of the plate. In this case, we view the Earth's surface from a lower altitude, so that the shape of this steeply dissected terrain is clear. Brightness in this image is strongly modulated by the topography, simulating dark shadows on the mountains. This kind of capability is relatively rare in geographic information systems, but common in computer-aided design and architectural graphics. We believe it is an important capability for spatial data processing, since the image results convey information to the user in a very efficient fashion.

Finally, we take the terrain surface, view it in a perspective presentation, and synthesize a dam in one of the canyons, filling the reservoir behind the dam with blue water (lower panel of Plate 3). Such a capability, merging GIS and computer-aided design functions, may be an important tool for many kinds of spatial analyses.

12.3 Waste Site Selection

Jensen and Christensen (1986) present an example of waste disposal site selection, based on a raster geographic information system (Plate 4). This practical example of the use of a GIS is a classic case of land-capability analysis. Their method begins by identifying the relevant constraints on locating a disposal facility. The constraints are based on regulations as well as on the transportation infrastructure required to support a waste disposal site. The initial criteria for siting a waste disposal facility, in their example, include:

- The site must be greater than 100 meters above sea level.
- The site must be more than 160 meters from a wetland.
- The site must be more than 160 meters from sensitive areas, such as endangered species habitats or recharge zones of aquifers.
- The site must be more than 200 meters away from any existing hazardous waste production facility or waste site.
- The site must be within 300 meters of a major road.

After determining these specific constraints, the needed environmental and cultural data are gathered. The required thematic layers include:

- locations of existing waste production facilities;
- locations of existing waste disposal sites;
- spatial extent of wetland land-cover types (cypress-tupelo swamp forest, bottomland hardwoods, etc.);
- spatial extent of upland land-cover types (mixed hardwood and pine, old fields, etc.);
- infrastructure (the transportation network, including primary and secondary roads and railroads, plus the utilities);

- locations of known sensitive areas (environmental reserves, endangered species habitats, archaeological sites); and

- elevation from a digital terrain model.

Many of these datasets are derived from stereo photointerpretation of 1:20000-scale aerial photography. The interpretations are transferred to a planimetric base map, and then digitized. The original themes are registered to a common map projection and scale, and are then converted into five separate intermediate layers (see Plate 4):

- a land-cover layer,

- a binary layer indicating the acceptable elevation range,

- a binary layer indicating areas acceptably far from wetlands and sensitive areas,

- a binary layer indicating areas acceptably far from existing waste production facilities and existing waste sites, and

- a binary layer indicating areas acceptably close to major roads.

A multiple-theme analysis of the datasets is then used to derive a new land-potential theme, indicating areas that meet all the necessary criteria. The analysis involves such analytic functions as proximity operators and Boolean operations for combining the intermediate layers to find all raster cells that meet all the initial criteria. Finally, a display and statistical analysis of the derived data layer is produced. The final map, indicating those areas in which it is feasible to locate a waste repository (Plate 4), can then be used to narrow the search for an appropriate property.

12.4 Irrigation and Water Resource Potential

Environmental Systems Research Institute (ESRI) is well-known for its ARC/INFO vector geographic information system. The United Nations Food and Agriculture Organization (FAO) contracted ESRI to develop a spatial database and resulting output products for estimating irrigation potentials for the African continent. We thank both ESRI and the FAO for the following material. The project was designed to identify areas that had sufficient water for irrigation, and those areas that would support agriculture if irrigated. The final map products were designed to support planning activities to try and avoid a repeat of the recent African droughts and famine.

The water availability and irrigation potential data were derived from a water balance model that was written by FAO hydrologists and programmed by ESRI. The water requirements of each country and each watershed were determined from soil, slope, and texture factors. Based on this data, the volume of water required for irrigation could be estimated. The next step was to estimate water resources. Surface-water resources were estimated from rainfall data, and groundwater resources estimated from geological and aquifer data. The model was then used to identify irrigable areas with a water surplus for each country and watershed.

Final map products show areas of irrigable soils and areas where rainfall makes irrigation unnecessary (Plate 5). Additional maps were produced to show areas with a water surplus. These maps were studied along with tabular listings of water volume and area of irrigable soils, stratified by country and watershed, to provide a first approximation of water supply versus needs.

This example shows a number of features of a complex spatial database creation and analysis project. The input datasets for this analysis are of variable quality, reflecting our relatively poor knowledge of soil and water in Africa. Data acquisition and preprocessing, for developing the digital spatial database, was a tremendous effort. The user's specific model of water balance provided the key analytic element of the project. And finally, the final products included maps of high cartographic quality as well as numerical tables.

12.5 Merging Raster and Vector Data for Map Update

Map update is an extremely common application of geographic information systems. GISs can provide an excellent set of tools for both the storage of maps of an area, and the updating of these maps when they no longer meet specified accuracy requirements.

In this example, which comes from northwest Atlanta, Georgia (Plate 6, provided by ERDAS, Inc.), the spatial database has two components. The image background is based on panchromatic data from SPOT, an operational satellite with a 10-meter pixel size. This data was rectified to the Universal Transverse Mercator coordinate system, using a ground-control point technique with a nearest neighbor algorithm (as discussed in section 6.6). Before displaying the image, it is enhanced with a high-pass filter (section 8.5), to highlight the cultural features in the image. The high-pass filter emphasizes boundaries between different types of land cover, and thus presents an enhanced image to the analyst.

The second data component in this example comes from digitizing a map. In this case, a 7.5-Minute quadrangle sheet was digitized, extracting a part of the transportation infrastructure. The railroad yard is portrayed in yellow, and roads are shown in blue and magenta. By displaying the transportation network (a fundamentally vector dataset) as an overlay on top of the satellite image, (a raster dataset), the analyst can rapidly focus on portions of the transportation network that have been constructed since the map was created (Plate 6). Similar analyses have been done to recognize and map the temporal change in land use in many areas.

In an operational system for updating maps, based on the ideas in this example, there are at least two different approaches. In one, the analyst uses the graphic display, which presents both the existing map data and the (presumably) new image data, to find features not represented on the map. Roadways that have been built since the map was last revised are easily identified in this manner. Next, the analyst essentially digitizes from the graphic display, to create the spatial information about the new roads, and thus, updates the data layers. In the other general method, automated techniques may be used to synthesize the road network based solely on the image data, and this layer is then compared with a similar layer based on the map. Disagreements between the two indicate changes in the features on the Earth's surface, which potentially provide the input for updating the map.

12.6 Species Habitat Analysis

An ongoing large-scale research and management effort involving several public agencies and private groups is attempting to restore the population of the highly endangered California condor (*Gymnogyps californianus*). The goal of this program is to establish a population of 100 birds (including 60 breeding adults) within the historic range of the species in California. A program of captive breeding has been instituted to increase the species population. The

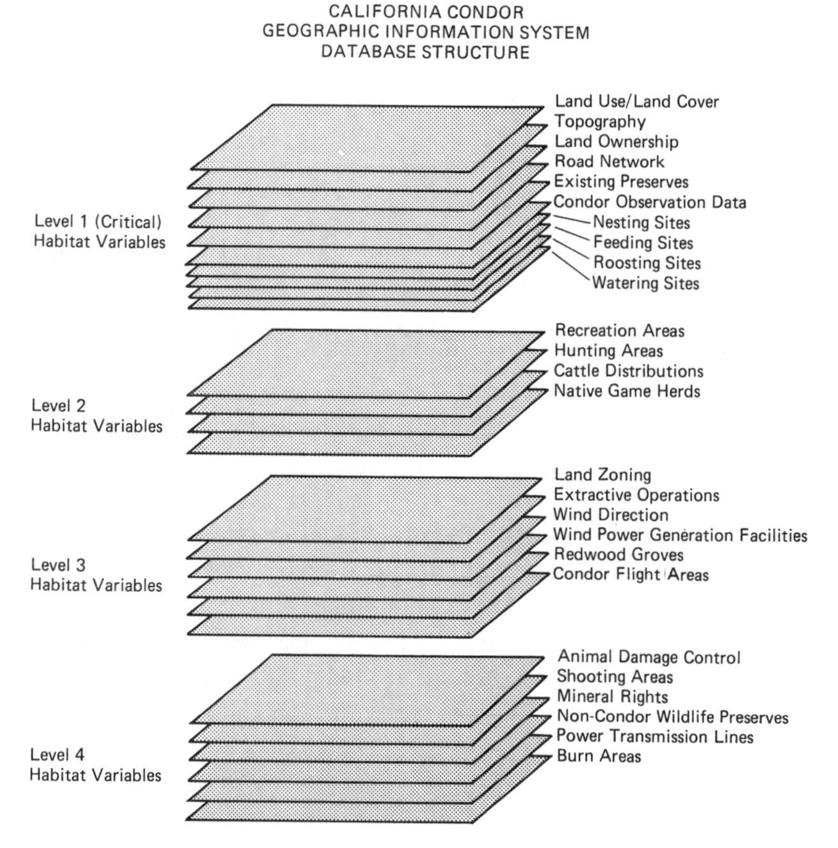

Figure 12.1 Condor Database Structure.

remaining condors are currently in captivity, and the breeding program has shown some early successes.

Environmental analysis of the historic range of the condor within California is an important component in the overall recovery program (Plate 7). The program is using geographic information systems to compile the necessary data for use in environmental analysis. This analysis involves combining condor observation data with selected environmental datasets, including land-use and land-cover and topography, to examine issues of map scale, resolution, classification, and accuracy (Figure 12.1). Projected use of this GIS for multispecies habitat analysis and species richness mapping is also evaluated. The program staff is using both raster and vector-based geographic information systems in the work.

Data layers which are currently included in the GIS include:

- A land-use and land-cover dataset (generated from manual interpretation of Landsat TM imagery and inputted to the database by coordinate digitizing);
- Topography in the form of digital elevation model (DEM) data acquired in standard 1-degree-by-1-degree blocks from the USGS; and
- Paved roads, approximate urban area boundaries, county boundaries, national forest areas and wildlife preserve areas. These themes were coordinate-digitized from the 1:250,000 clear plastic topographic sheets which are used as the base maps for the database.

The results of the initial work on the project demonstrate that a geographic information system is an efficient and effective tool for compiling and analyzing the large amount of disparate environmental data that is relevant to condor habitat. An interactive GIS is the only method currently available that can be used for predictive habitat modeling for the condor and for other rare, threatened, and endangered species within California.

12.7 Municipal GIS

San Bernardino Country, California, has developed a centralized facility for geographic information and processing. The following was kindly provided by Craig Gooch, manager of the Geographic Information Management System (GIMS), County of San Bernardino. A core group of GIS experts in GIMS, supported by staff from the County Office of Management Services, provides applications support, product generation, facilities management, and trouble shooting for a wide range of users. This centralized service is the result of a number of recent trends, including the need for improved productivity and service to the public, increasing demands for geographic information and analysis, and a desire to minimize redundancy and cost.

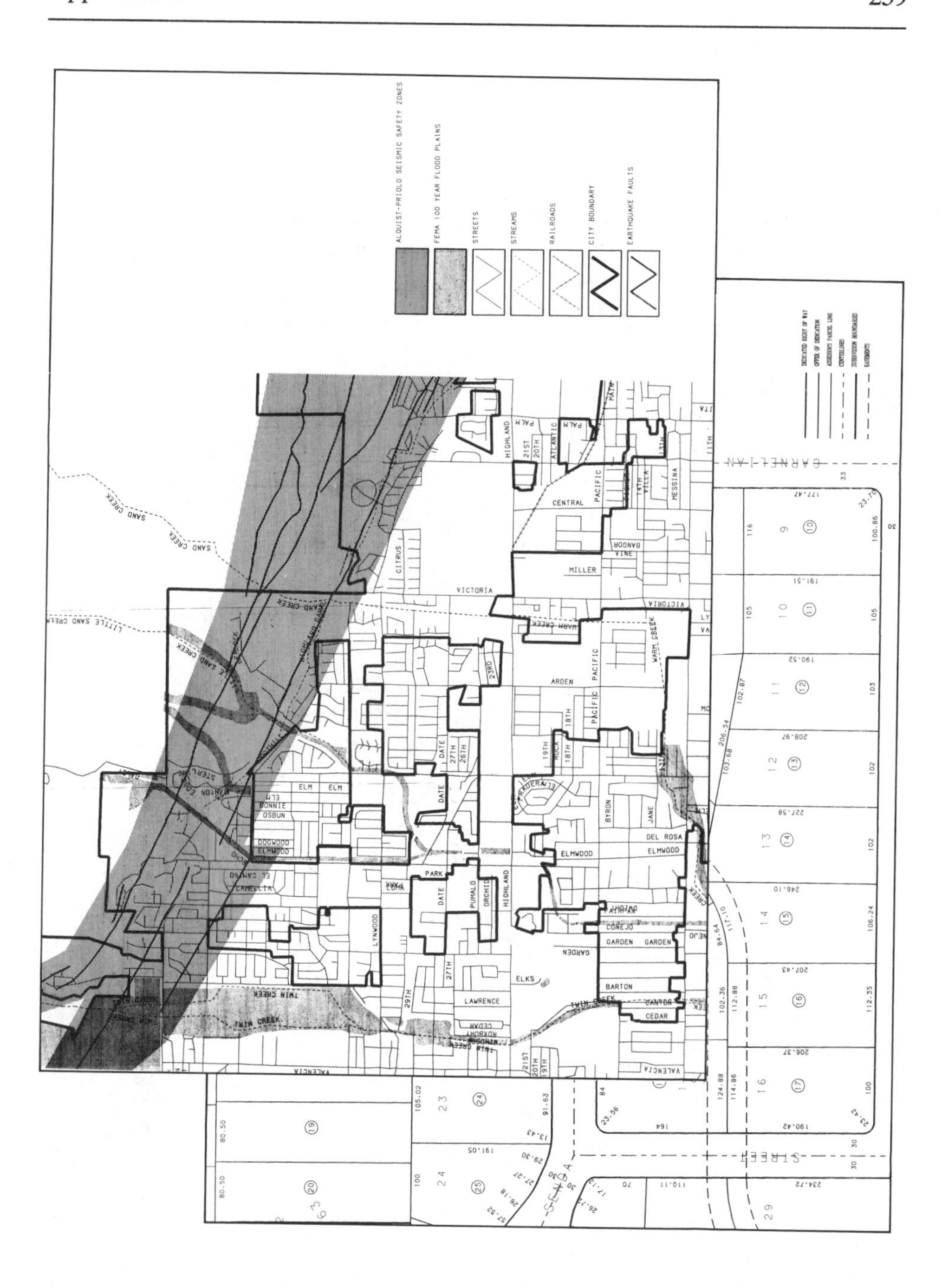

Figure 12.2 Centralized Municipal GIS.

The heart of the centralized repository of geographic information for the county is composed of three parts: survey control throughout the county, parcel boundaries, and a comprehensive street network. The primary survey network was developed from satellite positioning techniques, which are supplemented by photogrammetric methods and conventional survey records. The California State Plane coordinate system, based on the 1983 North American Datum, is the plane of reference for position information.

Manual assessor maps are the second element, and they have been digitized and corrected for errors. Standards are used so that placement and size of text as well as line symbology is consistent. Finally, the street network is stored as a set of street centerlines, with attached attributes to identify the street name, range of addresses for each street segment, and city and zip code. These two components, storing the parcel and street base for the county, form the basis for all subsequent GIS applications.

The GIMS users cover a wide range of municipal services and agencies. The Assessors office uses GIMS for parcel map production. The Registrar of Voters will use GIMS for voter and precinct management. County schools have active systems for analyzing attendance and developing pupil transportation plans. The county's Environmental Health agency has hazardous waste management and groundwater analysis applications. New applications are being developed for pavement management. In addition, a number of cities within the counties have cooperative data and analysis programs with GIMS.

To supplement these kinds of applications, a number of services are provided through GIMS, including applications development, training, system and user documentation, and database management (such as the development of access policies, development and administration of database standards, and quality control functions). Figure 12.2 shows examples of the some of the output products which are available from GIMS, including parcel maps and thematic maps showing flood and earthquake hazards.

12.8 Agricultural Production Modeling

The administrative bodies of Regione del Veneto in northern Italy has instituted a cooperative research program with the University of California. The goal of this program is to develop regional crop determination procedures and production and yield estimation techniques to support agricultural

PLATE 1
Color infrared aerial photograph. (Courtesy of author.)

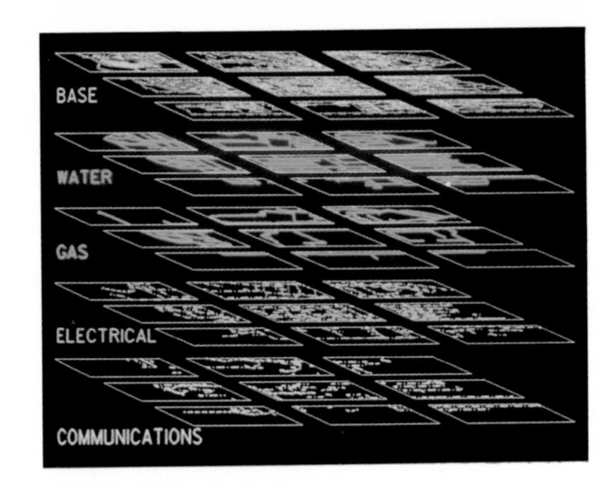

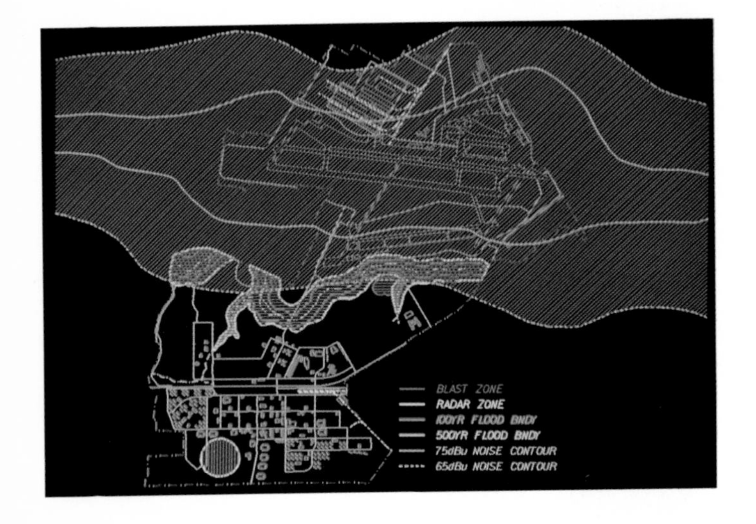

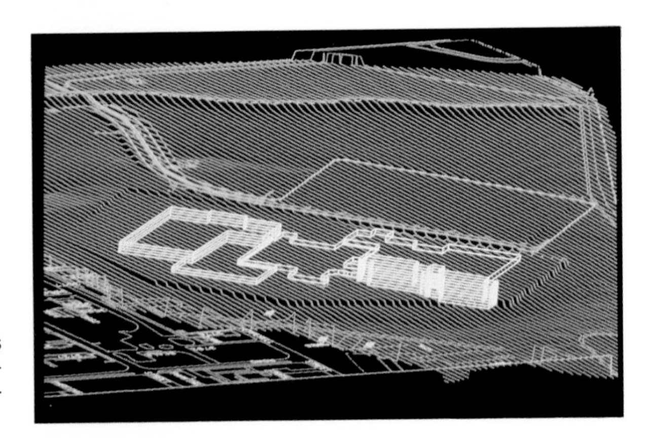

PLATE 2
Master planning. This graphic shows the data layers and zones of exclusion for a proposed hospital development. (Courtesy of Intergraph Corporation, Huntsville, Alabama.)

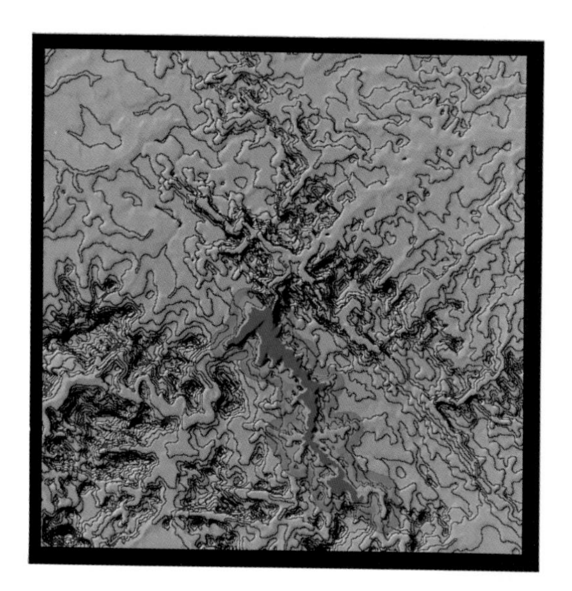

PLATE 3
Proposed dam site placement. This graphic demonstrates the use of three-dimensional graphics for site selection and analysis. (Image analysis by ERDAS, Inc., Atlanta, Georgia.)

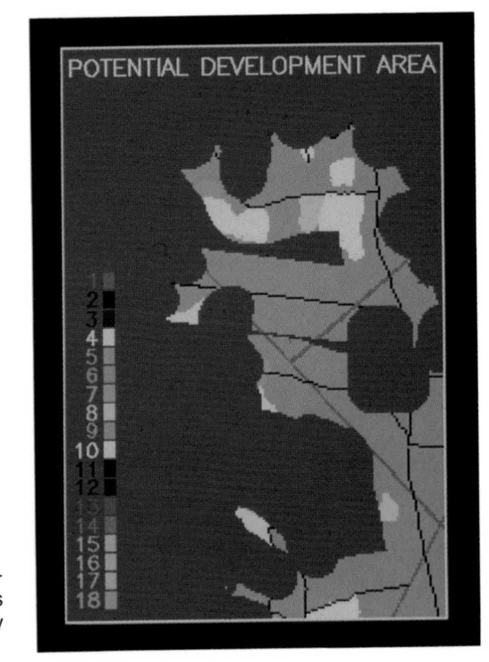

PLATE 4
Waste site selection. The important characteristics for selecting a site for waste disposal are joined with Boolean operators to create a map of appropriate locations for the site. (Courtesy of John R. Jensen.)

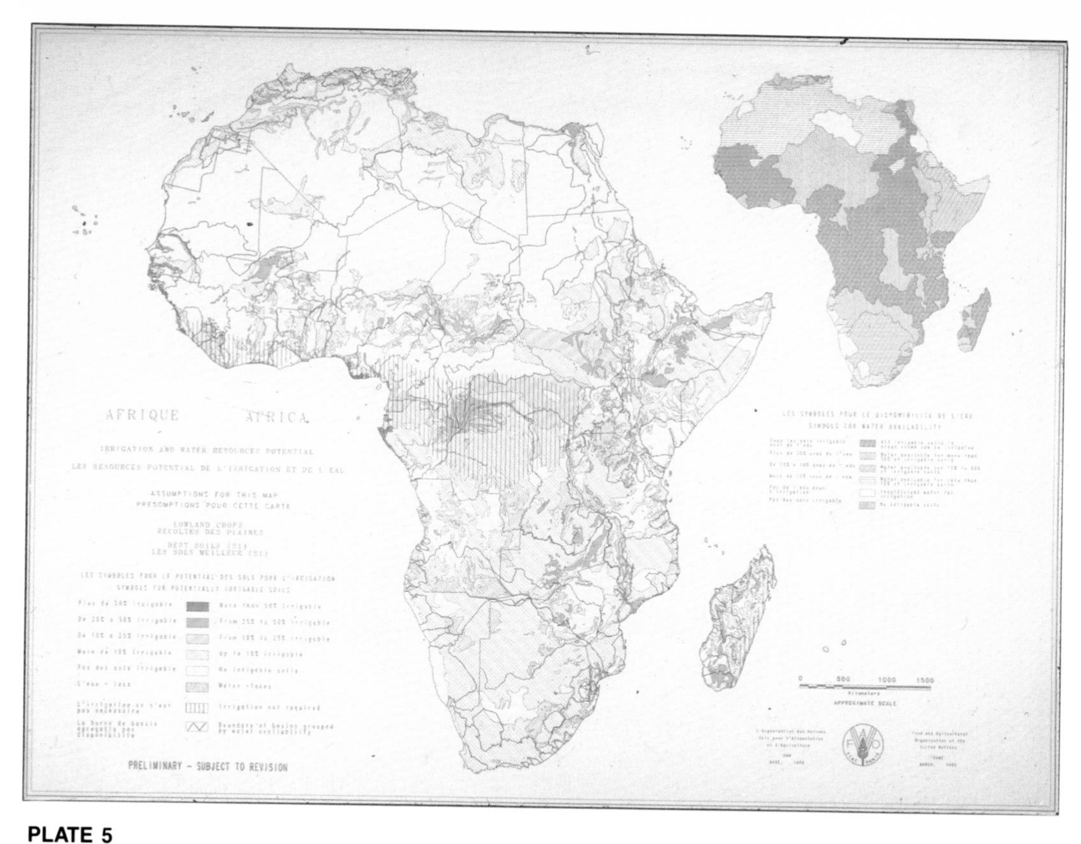

PLATE 5

African water resources. Many data sources were fused to create thematic maps of irrigation and water resource potential. (© Copyright 1989, Environmental Systems Research Institute, Inc. Used with permission.)

PLATE 6
Raster/vector data for map update. Raster datasets from satellite observations are merged to update vector map data. (Image analysis by ERDAS, Inc., Atlanta, Georgia.)

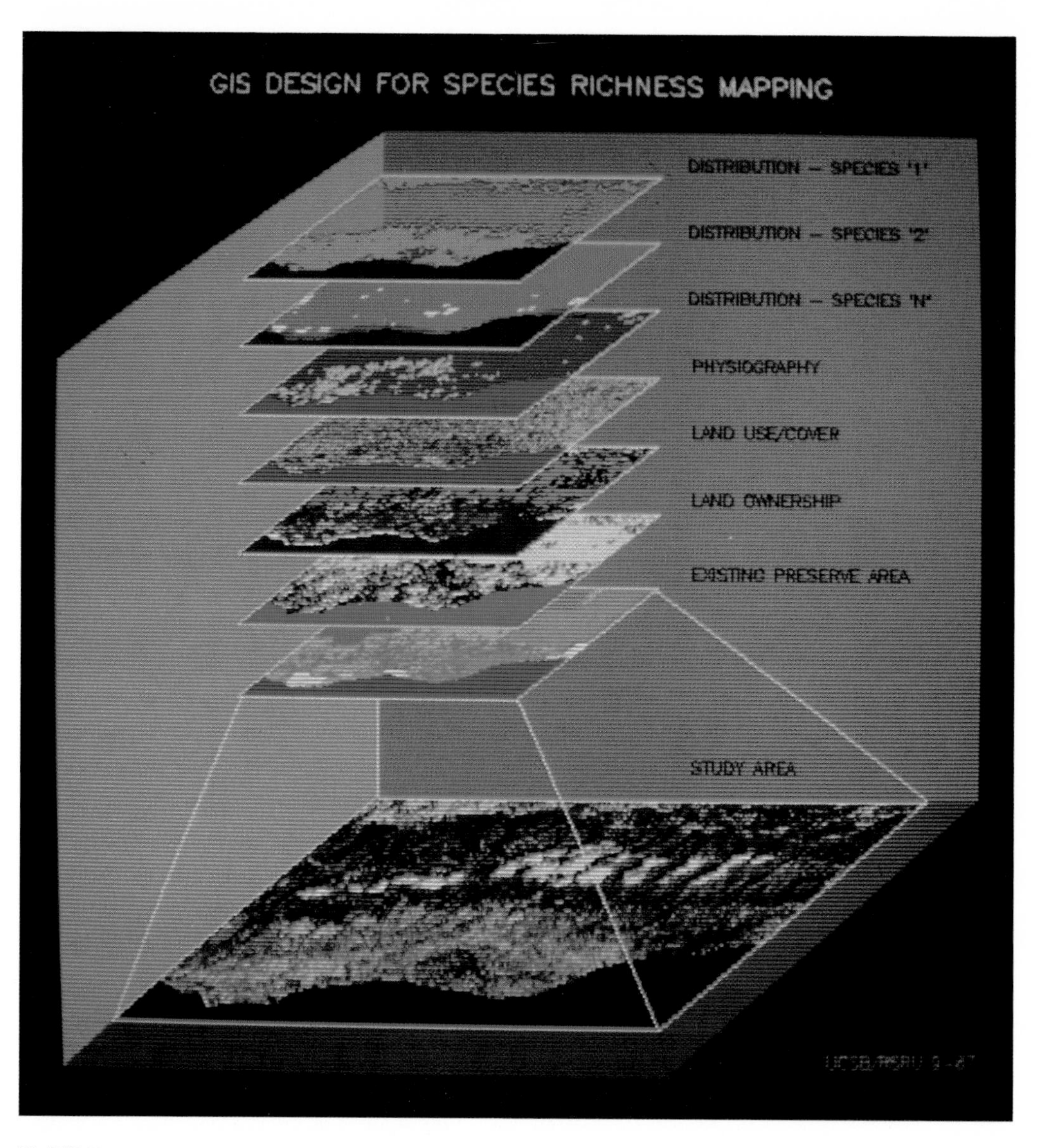

PLATE 7
California Condor database. This graphic comes from an analysis of observations of Condor activities as compared to site characteristics. (Courtesy of author.)

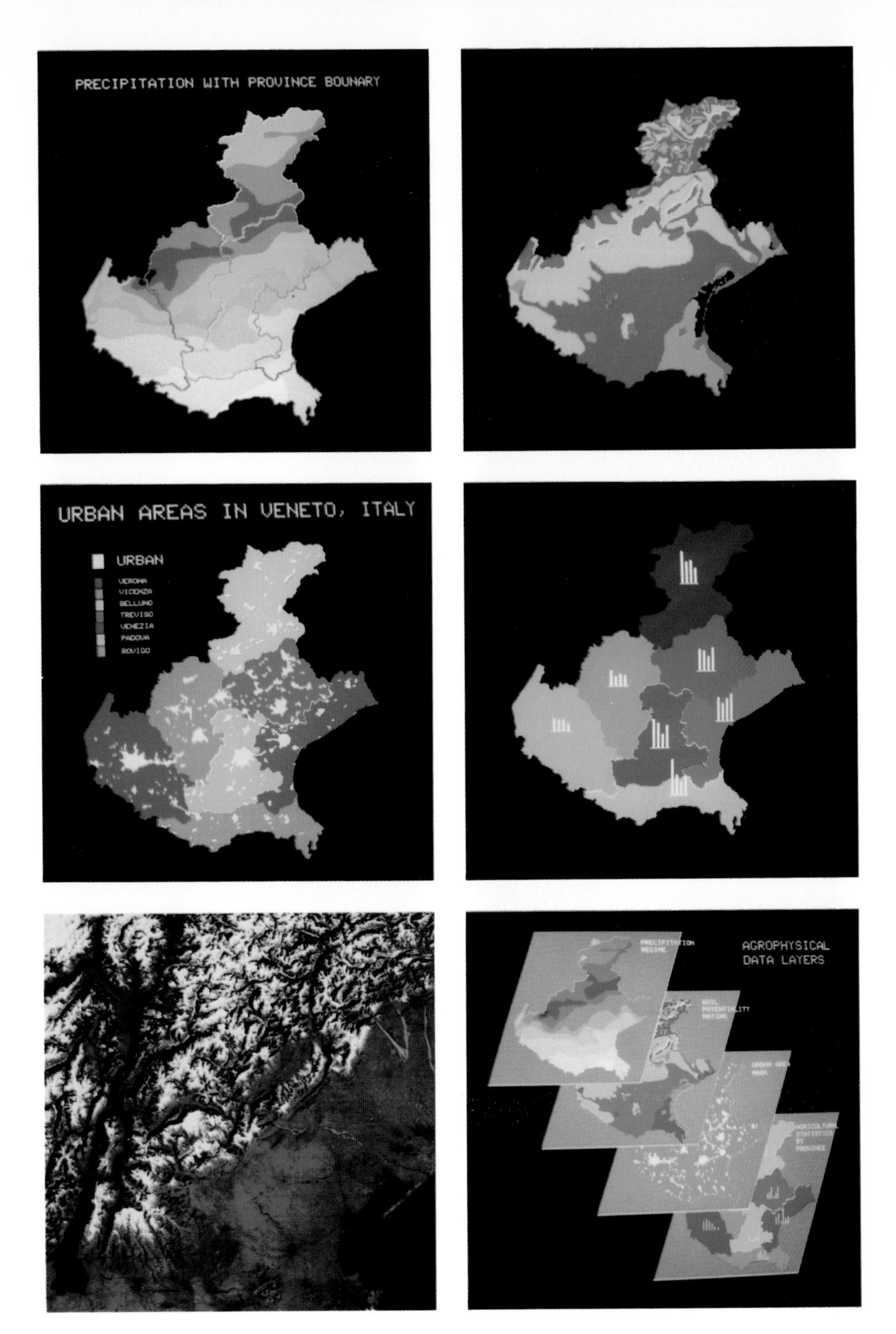

PLATE 8

Agricultural production modeling. Information about local characteristics, climate, and farming practices are used along with satellite observations to predict crop yield throughout the growing season. (Courtesy of author.)

resource management (Plate 8). This project is a part of an integrated environmental monitoring program being implemented for the Regione. Accurate crop yield prediction capabilities can increase the efficiency of agriculture in the Regione as well provide useful environmental data on the use of water resources and the use of chemical fertilizers and pesticides. Accurate, systematic crop inventory and yield and production estimation techniques are being developed for six principal crops grown within the Regione: small grains (wheat and barley), soybeans, sugar beets, corn, wine grapes, and orchard fruit.

Developing a system of crop inventory in this area presents some particular challenges. The Regione del Veneto contains an extremely complex mix of land uses and land covers. In addition, agricultural production in the Regione is characterized by small field sizes, a diverse crop mix, and multiple cropping in some areas.

A geographic information system format has been selected for use in compiling the large amount of environmental, economic, and agronomic data that is required for accurate crop production estimation and economic modeling. It is necessary to compile several layers of spatially referenced data for these purposes. These layers include:

- Crop classification and crop-class area maps. These will be generated yearly using multidate Landsat Thematic Mapper (TM) multispectral data and a suite of digital image-processing techniques.

- Meteorological data, including precipitation and temperature. This data is acquired in the Regione using a combination of digital and analog procedures. The data will be transferred from the Regione, reformatted (including coordinate digitizing if necessary), and inputted to the GIS. As some of this data is routinely received as point data (acquired by a group of irregularly distributed weather stations), processing is required to transform it into georeferenced area maps.

- Soils maps produced by the Regione.

- Maps of agricultural potentialities for various parts of the Regione del Veneto are produced by the Regione by combining soils, weather, and historic yield data.

Initial work in this study has demonstrated that both crop identification and crop yield modeling are data dependent. Accurate crop identification requires careful selection and timely acquisition of satellite data. Agrometeorological-based crop yield modeling is dependent upon complete and accurate historical yield and weather datasets. At this time, several efforts are proceeding in parallel. Economists are building databases of crop sales and prices. Agronomists are monitoring a series of test sites in the region to provide ground truth for crop practices such as planting and harvest dates and irrigation practices. Remote sensing specialists are developing numerical models of crop growth, to be able to identify crop types and predict yield at the end of the season based on multi-temporal satellite observations. A raster-based geographic information system forms the core of the analysis software.

12.9 KBGIS: Artificial Intelligence and Geographic Information Systems

The development of a geographic information system based on principles of artificial intelligence and data structures has been a part of ongoing research activities at the University of California, Santa Barbara. The system is called the *knowledge based geographic information system*, or KBGIS. The essence of the system is described in Smith et al. (1987b). We will briefly outline the key components of the system and will describe one or two applications.

KBGIS contains two principal databases. One, the **object database**, contains information about discrete spatial objects. It is implemented by means of *frames*, a powerful structure that allows for inheritance. For example, we may define the object *well*, which has among its attributes location and depth. When we later define a new object *water well* as a particular class of *wells*, it "inherits" characteristics of the parent. This can be a useful way to structure knowledge about spatial objects.

The second database in KBGIS is a **quad-tree**, as discussed in Chapter 4. Separate quad-trees are maintained for each thematic layer in the database. Continuous data, such as elevation, may be kept in nodes of a tree, as may categorical data, such as classes of land use. For continuous data, at coarse resolution levels of the tree, minimum and maximum values in addition to mean values may be stored, aiding search. For categorical data, at higher levels of the quad-tree, the complete list of all categories included at the relevant smaller cells is maintained.

The system characterizes spatial objects in terms of three classes. The first class consists of pixel properties, that is, the kinds of properties stored at a single pixel in an equivalent raster GIS data layer. These include such things as the land-cover class at a location, or the elevation there. The second class of properties consists of groups of pixels. The most natural of these group properties are such things as the area, perimeter, and shape of a homogeneous area. The third class of properties involves relationships between areas. This final class includes the distance between two areas, or the direction, relative to true north, of the line connecting two areas at their closest approach.

The system is able to learn in several ways. New object definitions may be created by explicitly teaching the new object definition to the system, via a menu system. For example, the object "good place for an airport" might be defined as complex object with the following characteristics:

- a contiguous area of greater than a specified size,
- an area having a certain class of permissible land uses,
- an area located within a certain distance of a principal road, and
- an area within a predefined distance of a population center.

Of course, the relevant data layers must be loaded into the system for the system to be able to use such a definition. The definition is created at one time, and then the system may be told to **learn** examples of such a complex object by searching the quad-tree database.

Search may be guided heuristically, to minimize the costs of search. In the example above, the system can exclude all areas in the database that are not within a specified annular region around a population center, and thus can avoid searching all of the spatial data. Such a capability can make tremendous differences in the performance of a geographic information system.

Inductive learning is also implemented in the system. By inductive learning, we mean that the system can determine the general characteristics of a class of objects by studying a number of specific examples. In KBGIS, the analyst enters coordinate locations that correspond to places of interest. The system then characterizes each location in terms of the information it has about these places. The key step occurs when the system generalizes a new object description from these known examples, and the description is consistent with the examples. Finally, the system can be asked to locate new examples of the object.

As an illustration of the inductive learning capability, the U.S. Geological Survey provided a raster dataset over the Black Hills of South Dakota and Wyoming. Layers in the dataset included surficial geology, elevation, slope, and categories of land cover. The geologists specified four coordinate locations in the dataset and a spatial window around these locations to consider as part of the site. The system was instructed to consider the four data layers above, as well as the derived characteristics of size of spatial objects, distance between objects, direction of one object from another, and containment.

After determining the specified characteristics of the spatial objects at the four sites, the system generalized the properties at the sites to determine the common characteristics of the four. In this specific example, the generalized description derived by the system was based on three spatial objects. In the vocabulary we have used in this text, these spatial objects would be (possibly complex) polygons. One object was characterized by its land cover of forests, while the other two were characterized by their surficial geologies. The system also determined size constraints on the three objects: the forest-covered region covered the entire window around the site, while the other two objects had derived minimum and maximum areas. Finally, the system determined that the two polygons characterized by their surface geology were oriented east/west of each other, and that the forest-covered region contained one of the polygons defined by surface geology (in other words, the boundaries of one were contained within the boundary of the other). This complex object description is automatically added to the system's knowledge base, and the system can then be asked to find other examples of this new object.

Chapter 13

Looking Toward the Future

Applications of geographic information systems are increasing rapidly, as city, regional and environmental planners, resource managers, and the scientific community become aware of the full potential of these systems. Both science-oriented and applied-oriented users of GISs are learning that these systems can help them in many ways. They assist in focusing user's data-acquisition activities, provide a framework for improving data storage, and provide tools that can facilitate data management. GISs also provide these users with analytic and reporting capabilities (including graphic production tools) that were unheard of 20 years ago. Systems outputs in text, tabular, and graphic forms can be used to present the results of research and operational programs in a positive, easy to understand form.

Most users involved in geographic analyses want to see a map that contains information relevant to their application. Along with significant federal and state requirements for thematic products, the business community has definite thematic mapping requirements. In particular, private concerns that supply information derived from remotely sensed data have found that almost all client requirements are, to a large extent, cartographic in nature (Simonett, 1976).

To a very real extent, map accuracy is an important and unresolved issue. Assessing the accuracy of thematic maps is a difficult issue in its own right. A number of publications deal specifically with this subject (Rosenfeld et al., 1981; Rosenfeld, 1982; and Estes et al., 1984, among others). The problem is compounded in GIS applications, where data layers with differing accuracies are combined to produce a new product. Under very limited circumstances, it is possible to derive the accuracy of a resulting data layer from statements of accuracy in the original input data. In the vast majority of cases, we cannot ascertain the accuracy of the resulting data layer.

For many practical reasons, the data input to a geographic information system are simply the "best available", particularly when dealing with thematic maps as source data. The way in which errors propagate through a spatial data analysis task is an important research topic. In addition, because monitoring inherently involves change detection, and because modeling can involve the analysis of processes that can add their own variance to data, a user can readily see the importance of knowing the accuracy associated with various GIS data inputs.

Monitoring is central to environmental planning and management applications. It is of direct interest not only to federal agencies, such as the Environmental Protection Agency in the United States, but to local agencies as well. This is also true on the international level, as we move towards the era of the International Council of Scientific Unions Global Change Program, and NASA's Mission to Planet Earth. Monitoring is also central to a variety of other applications in such areas as agriculture, hydrology, and urban and regional analysis. Modeling has become increasingly important as computer technology has advanced. Advances in processing power and techniques have facilitated the use of models for the analysis of a variety of spatially distributed phenomena, from atmospheric and ocean circulation to the migration of materials through soil and groundwater.

An example of the operational use of geographic information system technology is found in a paper by Tinney and Ezra (1986). The authors bridge from use of GIS technology in the Department of Energy's Comprehensive Integrated Remote Sensing Program to a discussion of specific programmatic goals. These programmatic goals include the production of reports, maps, charts, and other visuals and computer generated products. Datasets produced by this program may be used in a wide variety of studies from site development planning to emergency response and disaster control planning. One major emphasis in this paper is the use of remotely sensed data for providing input to the Department of Energy databases.

Many of the difficult and expensive problems arising in the design and use of a GIS come from the very large and rapidly increasing volumes of data that users demand. Resource managers, scientists, city planners, and other users of digital geographic data are acquiring information from a broad range of sources. Both practical and theoretical problems develop when users attempt to integrate these datasets.

More data is also becoming available in digital formats. This will help to expand the use of GIS technology, by reducing the cost of data access and preprocessing. Agencies such as the Departments of the Interior and Commerce in the United States, as well as international organizations such as the United Nations GEMS (Global Environmental Monitoring System) and UNEP/GRID (United Nations Environment Program Global Resources Information Database) programs continue to increase the volume of their annual output of data in coded digital formats. With this increase in the volume of digital data, from a wide variety of agencies around the world, the issue of standard data formats becomes important.

The National Aeronautics and Space Administration has been a significant provider of digital datasets. Satellite digital data volume is projected to increase dramatically in the late 1990s with the launch of the Earth Observing Satellite (EOS) system. This planned polar orbiting system has projected data production volumes on the order of 1 terabyte per day. To illustrate this incredible data volume, 1 terabyte of digital data, if stored on standard IBM-format 5.25-inch floppy disks, would require 2.7 million disks, and require a shelf almost 7 kilometers long. Alternatively, using standard 1600 BPI magnetic tapes, hung on ordinary tape racks, EOS would fill a room of 900 square feet each day, for the 40,000 magnetic tapes the system may produce. Intelligent acquisition, selective storage, manipulation, and analysis of these data will require advanced geographic information systems if we are not to drown in this sea of environmental data.

NASA is of course not the only U.S. Federal agency that is creating large volumes of digital data, and is interested in facilitating the use of these datasets. At this time, an Interagency Working Group for Data on Global Change has been established within the government. One of the objectives of this group is to plan for a virtual national database by the mid-1990s. Such a data system has important implications for both current and future users of GIS technology. The implementation of such a system would provide a single point of contact for such diverse spatial datasets as demographic data from the Department of Commerce, bathymetric data from the National Oceanic and Atmospheric Administration, topographic data from the U.S. Geological Survey, and remotely sensed data from NASA sensors. Research will be needed in a number of areas to be able to develop such a system, including heterogeneous database management, network communications, security, and distributed processing.

Computer and geoprocessing technologies that support GIS operations are evolving to higher levels of capability. Computer technology, particularly in terms of hardware for data processing and storage, is advancing at a tremendous pace. New algorithms and data structures are being developed. Overall, we have moved from an environment in which only large institutions could support GIS activities to one in which researchers and resource managers can have useful analytic, data storage, and display capabilities in the personal computer on their desk. In addition to advances in traditional computer technology, research groups around the world are looking at fields of artificial intelligence (e.g., expert systems, natural language understanding and image understanding) to make geographic information systems more efficient and more friendly to users with limited computer literacy.

Based on our work with geographic information systems, we see that growing knowledge and practical experience is providing greater capabilities, at the expense of greater complexity. This directly affects the costs of software (in terms of programming for new functions), database development, and maintenance. This complexity must be balanced against preserving flexibility to being able to incorporate new analytic models as well as new hardware over time. The underlying data models and structures in a practical system determine the efficiency of different processing functions. In addition, with the cost of data-processing systems generally decreasing (or from another perspective, the price/performance ratio decreasing), new applications are becoming economically feasible.

For those still looking for aid in selecting and implementing a GIS, Streich (1986) will be helpful. Streich states that with the basic GIS procurement process becoming more routine, environmental resource managers can pay more attention to customizing GISs to solve specific applications problems. He discusses potential problems that may arise when bringing a digital GIS into an organization. He concludes that with proper planning and good communication channels with institutions engaged in similar areas, management personnel charged with environmental protection responsibilities will be able to maximize their return on GIS technology investment. Indeed, papers such as this by Streich and another by Star, Cosentino and Foresman (1984) should be required reading for prospective developers and purchasers of GISs.

Another important lesson to remember is that as a GIS becomes an established element of the way an organization conducts its operations, it often happens that the system is found by a wider user community, with broader ranges of needs, than originally intended. As with the example of the Canadian Geographic Information System discussed in Chapter 2, care must be taken that systems do not become overwhelmed with users and find that they cannot adequately respond to the original requirements. In a sense, this problem is no different for the managers of any other data-processing or information system. Management must also face the continuous need for technology upgrades. Hardware and software improvements through time must be a specific part of system life-cycle management practice. These last comments imply that geographic information systems are not static.

By way of further general observations, we see that there are still few commercial vendors marketing fully integrated geographic information systems. The reasons for this are not all that clear. The overall complexity of the development of such systems is of a course a hindrance. A perceived lack of a current market that is sufficient to support a major development effort is also a key factor. The domain, discipline, and application-specific nature of many GISs tends to encourage the development of in-house systems, and this may also play a role. The presence of established commercials firms that already have a large market share and already provide user support in the form of systematic upgrades and documentation and training is yet another possible inhibiting factor.

Finally, the practical design and application of digital geographic information systems is a young field, and as we hope we have shown, an interdisciplinary one. Smith et al. (1987a) discuss what should be required of a GIS. A GIS should:

- be able to work with large, heterogeneous spatial databases;
- be able to query the databases about the existence, location, and characteristics of a wide variety of objects;
- operate efficiently, so that the user can work interactively with the underlying data and the required data analysis models;
- be easy to tailor to a variety of applications, as well as to many kinds of users;
- be able to "learn" in significant ways about the data and the user's objectives; and
- be able to supply a readily interpretable output product for the ultimate users of the system.

We look on this list as an entry into research topics rich in complexity and interest for a wide range of disciplines, as a set of objectives that a user might try to levee on a system's developers, and as a set of goals that a developer may strive to attain.

Figure 13.1 The upper figure shows a three-dimensional representation of the South American Continent, with clouds above the land surface (Courtesy J. Hall and B. Mortenson, Image Processing Laboratory, Jet Propulsion Laboratory). The lower figure is an enhanced image from a Photon Scanning Tunneling Microscope (Courtesy Biomedical and Environmental Sciences Division, Oak Ridge National Laboratory).

As we mentioned in section 1.6, scientists are interested in objects and phenomena at both the smallest and largest of scales. The images in Figure 13.1 provide a good example of these extremes. The upper part of the figure is a composite of two three-dimensional databases, to form this perspective representation looking north from Antarctica to the South American continent. The cloud data, which depicts the base of the cloud layer, is from the High Resolution Infrared Sounder on the NOAA 7 satellite. The terrain information is derived from ten-minute resolution digital elevation data from the National Center for Atmospheric Research. Total relief in this dataset is on the order of 20,000 meters. The lower part of the figure is a computer-enhanced view of a quartz microscope slide, taken by the Photon Scanning Tunneling Microscope. The mountainous structure is a residual of the polishing process; total relief in this dataset is only 200 nm. These images illustrate an important issue presented in this text: the search to discover new methods to visualize environmental elements and processes in space, from micro to macro scales.

Finally, geographic information systems technology is one more way in which we can improve our understanding of the Earth. Much recent discussion in the United States has emphasized our general lack of geographic understanding. From our admittedly biased point of view, we are convinced that an understanding of geography, in general, and processes distributed in space, in particular, is important in today's society. It is our hope that this text is a small step in the process of realizing the full potential of geographic information systems.

Suggestions for Exercises

The follow are suggestions for classroom exercises. Some require access to an automated geographic information or image-processing system, while many others do not. Some require nothing more than paper and a pencil. This list is cross-referenced to the relevant chapter.

Chapter 2: Background and History

A. Inquire of your regional natural resource, forestry, or water resource agency:
What spatial datasets are kept? How are they storing this data? What base map projections and map scales are in common usage? Is there a systematic means of updating their spatial data? How long have they been doing it this way?

B. On a timeline, indicate the principal dates and accomplishments in the development of GISs. Then find key dates for the computer industry (for example, IBM's introduction of the System 360) and space exploration (for example, the U.S. Landsat program), and enter these dates and accomplishments as well.

Chapter 3: The Essential Elements of a GIS: An Overview

A. Contrast the five elements of a Geographic Information System with the equivalent elements of an employee records management system. Where are the similarities in staff roles, in hardware, and in software?

Chapter 4: Data Structures

A. With paper and pencil, roughly enlarge a local street map to a scale of approximately 1:1000.

(a) Develop a raster dataset of the transportation network: On a light table, lay a sheet of millimeter graph paper over the enlarged street map, and mark off the cells corresponding to the center lines of the principal streets.

(b) Develop a vector dataset of the transportation network: Lay a second sheet of millimeter graph paper over the map and draw the center lines of the roadway as a series of straight line segments. Use the cells at the end points of these straight vectors to indicate the points that might be termed nodes in a vector database.

(c) Compare the number of raster cells coded in (a) with the number of end cells required in (b).

B. Examine the arc-node data structure example in this chapter (Figure 4.12).

(a) How would you revise this example so that *all* the information in this example is stored in a raster data structure?

(b) Estimate the relative data volume of the arc-node example in comparison with your choice of raster representation.

C. If a relational or flat-file database management system is available, develop a single database which stores spatial objects. An example might be based on countries in Europe, with country name, population, and the x-y coordinates of the capital city stored in the database. Compare this simple system to the vector structures discussed in the chapter, in terms of the kinds of analysis that may be possible.

Chapter 5: Data Acquisition

A. When was the last time your national government updated the maps of your local area? (in the United States, for example, the relevant map would be a USGS 7.5-minute topographic quad sheet.)

B. Contact your local municipal planning agencies. When was the last time they revised the local land-use plans? Are aerial photographs a part of the revision process? Are automated systems in use?

C. Find an aerial photograph of a local housing development, preferably at a relatively large scale.

(a) Choose a set of land-cover classes that represent the majority of the area in this development (such as residences, gardens, roads, recreational areas, and so forth).

(b) Sample the data in the photograph by choosing random locations in the photo (via a random number table, or a computer program) and noting the apparent land-cover class at each location. Keep track of the proportions of the different classes as you increase the number of samples.

(c) How many samples seem necessary for a stable estimate of the fractional cover of the dominant land-cover classes? A plot of fractional cover versus the number of points is a good way to display the data.

(NOTE: If there are a large number of students, this exercise may be efficiently run in small groups, with the groups comparing answers at the end of the exercise.)

Chapter 6: Preprocessing

A. Programmers:

(a) Try coding a vector to raster conversion routine: read in an integer to indicate the number of points in the vector, followed by the sequence of formatted x-y data value pairs, and convert the data into a raster of specified size and scale. For an output raster of the order of a few hundred cells on a side, use an ordinary memory-based data array.

(b) What strategies might you use when the output raster is too large to contain in memory at once?

B. Everybody:

(a) Using an ordinary photocopy machine, duplicate two adjacent maps from a well-known small-scale map series (1:250000, for example). Mechanically match the edges as well as you can, and tape the edges together.

(b) For ten of the features that span the boundary (such as contour lines, roads, water bodies and so forth), estimate the worst mismatch of the features along the boundary, in both millimeters on the map and in meters on the ground

(c) Try and quantify the size of the errors you have created at the boundary by calculating the *Root Mean Square Error*: estimate the mismatch, in terms of meters on the ground; then calculate the RMS error as follows:

$$\text{RMS Error} = \sqrt{\sum \text{of squared errors}/\text{number of measurements}}$$

C. Class project:

(a) Everybody together: from a map, determine the geodetic coordinates of a set of ten unambiguous locations: benchmarks, roadway intersections, etc. Write down the (northing-easting) tuples.

(b) Everybody separately: Set up a raster data array, each choosing different starting coordinates and cell sizes. Convert the point locations from the map to the relevant cell. Write down the list of cell locations of the objects, and convert these cell centers back to geodetic locations. Calculate the RMS error. This shows you the error inherent in converting between data structures.

(c) Everybody together: graph RMS error versus cell size. What patterns do you see?

D. If a GIS or image processing system is available:

(a) Take a small raster dataset with integer categories, such as a 100-by-100-cell array of land-cover classes, and rotate the orientation of the array clockwise by 30 degrees.

(b) Then, rotate the results counterclockwise back to the original orientation. Compare the results when using a nearest-neighbor algorithm versus a cubic convolution.

(c) Visually, how does the final dataset compare to the original? Numerically, how would you use the tools of the GIS to compare the original data to the rotated-and-then-counter-rotated data?

E. Interpolation: Obtain large scale (approximately 1:10000) topographic maps, that show areas of low relief and areas of high relief. For one low relief area and one high relief area:

(a) Locate five points, and estimate their elevations.
(b) Select a point which is in the midst of the others. Call the selected point the *test* point, and call the others *known* points.
(c) Average the elevation of the four *known* points; compare this simplest value to the elevation of the *test* point estimated from the map.
(d) Select two *known* points, that are roughly on either side of the *test* point. Using the map scale bar, estimate the distance between the *test* point and the selected *known* points.
(c) Add the distances between the *test* point and the two selected points together; call this sum S.
(d) For the two selected points, weight their elevation by (S minus the distance to the *test* point) / S).
(e) Add the two weighted elevations, to compute an interpolated elevation for the *test* point. Compare the resulting value to the elevation as estimated from the map.
(f) For another comparison, consider fitting a surface to the *known* data points to predict the *test* point. Use a statistics program to fit either a multiple linear regression or a second-order polynomial regression, to predict elevation as a function of x and y position. Use the *known* data points to calibrate the model, and then test its performance with the *test* point.

F. Digitizing reproducibility: A standard problem in the preprocessing phase is understanding the errors in the process of digitizing. If a GIS or CAD/CAM system is available, each student in the class may digitize a small set of features. Then each student should determine the area and perimeter of the features, the lengths of a few lines, and the relative positions of a pair of points (both displacement and direction). As a class project, plot the histograms of each of the measurements to develop a feeling for the reproducibility of the process.

Chapter 7: Data Management

A. With a ruler, measure the size of a box of ten standard personal computer floppy disks, an optical disk (a stereo CD is fine) in its case, and a standard 9-track magnetic tape. Find out the amount of data which may be stored on the floppy disks. Assume that the 9-track tape is 2400 feet long, with a useful linear data density of 2 kilobytes per inch. Assume that the optical disk holds 400 megabytes of data. Compare the volumetric data storage efficiency of the three storage mechanisms.

B. Inspect a copy of your local telephone directory.

(a) Estimate the number of bytes of data in the directory by noting the number of pages, the number of lines of text per page on a randomly selected page, and the number of characters per line from a few randomly selected lines. Assume 1 character on the page will require 1 byte for storage.

(b) How many floppy disks will it take to store the telephone directory? How many optical disks? How many magnetic tapes?

C. Find out the area of the county or municipality in which you live. Assume that we wish to store data about this geographic object in an uncompressed raster data structure with a 1-meter cell size. The themes to consider are elevation (requiring 4 bytes of storage per cell), plus one more byte each for land-use classes, population density classes, and municipal services codes. How many floppy disks would be required to store this data? How many magnetic tapes? How many optical disks?

D. Projections for NASA's Earth observing space platforms in the late 1990s suggest a data-production capability of approximately two terabytes of data per day. Calculate the length of the shelf required to hold a single day's data on floppy disks. Assume that magnetic tapes are stored reasonably efficiently, how many meters of wall space, 3 meters high, will be required to hold 1 year's data? How many meters of wall space will be required for optical disks?

Chapter 8: Manipulation and Analysis

A. Why doesn't it make sense to use a mean spatial filter on a nominal dataset? Why does it make sense to use a median filter on such data, at least in some circumstances?

B. Thematic data overlay:

(a) Find a map of a local park or recreation area, with a map scale of approximately 1:5000 or 1:10000.

(b) Lay a sheet of plastic over the map and sketch the trails on the plastic.

(c) Sketch an impact boundary area around the trails on the plastic sheet, the impact area having a total width of 150 meters.

(d) Estimate the total area affected by foot traffic on the trails, as sketched in part (c), in several ways:

- using a random dot grid,
- using an electronic planimeter, and
- cutting the area out of the sheet and weighing it on a precision balance.

(e) Compare your area estimates, and comment on sources of errors in each of the estimation techniques.

C. Examine one of the issues in interpolation.

(a) Draw a 5-by-5 cell grid on a sheet of paper; number the cells 1 to 5 in the x and y directions.

(b) Calculate the value of each grid cell -- call it Z -- by the following function:

$$Z = x/5 + \sin(y/2)$$

where y is in radians.

(c) Choose five random locations in the gridded area and estimate their locations to 1/10 of a grid unit. Either use a random number table, or a computer, or (literally) throw darts at the grid.

(d) Using a nearest-neighbor rule, select the value from grid cell whose center is closest to each random location. Compare this value to one calculated from the equation above.

(e) Subdivide one of the 5-by-5 grid cells into four equal smaller cells. Devise a rule for estimating the values of the four quadrants from the original large cell values, and compare them to the values calculated from the equation above.

Chapter 9: Product Generation

A. How fine a line can you create, using either a technical pen or computer plotter? Investigate the capabilities of the computer printers and plotting devices in terms of lines (or dots) per unit length and the total number of available colors or gray levels.

B. Estimate the resolving power of your eye:

(a) On a sheet of paper, draw a set of six parallel lines 4 millimeters apart, approximately 20 to 30 mm long.

(b) Hold the page at different distances from your eyes and estimate the distance at which you are unable to reliably distinguish the individual lines.

(c) Based on this distance, calculate the minimum useful line spacing in terms of the angle subtended from one line, back to your eye, to the next line.

(d) What does this tell you about a useful raster display size on a 50-by-50 centimeter sheet of paper at normal viewing distances?

C. At your local computer center, compare costs of line printer plot versus the available plotters, versus a color-film hardcopy device if available. Include in this survey the number of unique colors possible, the available sizes of finished product, and the amount of clock time required to produce an average graphic.

D. Aliased lines:

(a) On a sheet of centimeter graph paper, draw two straight lines: one at approximately 45 degrees from the vertical, the other approximately 10 degrees from the vertical.

(b) Darken the cells that either line crosses: thus we have converted these vectors to a raster representation, with no anti-aliasing.

(c) On a second sheet of paper, draw the same vectors, but try and draw an anti-aliased raster, based on the discussion in Chapter 9.

(d) Attach each of these sketches to a wall. Carefully look at these two sketches from various distances from the wall, and comment on the quality of the raster-converted vectors as a function of viewing distance.

Chapter 10: Remote Sensing and GIS

A. Find out what air photographic coverage, at what scales, is available for your local area.

B. If a geographic information or image processing system is available:

(a) Choose a single-channel raster image.

(b) Rectify the original image to a different scale and orientation, and then rectify the image back to its original scale and orientation. For a class project, arrange that each of nearest-neighbor, bilinear, and cubic interpolations are all used for comparison.

(c) Compare the original image to the doubly-rectified image, on a pixel-by-pixel basis, to try to detect change.

(d) Alternatively, classify the original and doubly-rectified images separately, and perform a symbolic change detection. Comment on what differences you found in comparison to Step c.

C. If time and facilities permit, demonstrate the concept of multispectral imagery:

(a) Mount a camera on a tripod overlooking a complex scene that includes vegetation, roads, buildings, clouds, etc. An oblique view from a window in a tall building is often good for this purpose.

(b) Take a number of exposures of panchromatic film through a variety of different filters -- haze correcting filters, primary and secondary colors. If an infrared emulsion is available, expose this as well according to manufacturer specifications.

(c) Print the various exposures, and try to develop spectral signatures for a variety of spatial objects in the scene.

(d) For several types of objects, can you suggest an minimum set of spectral bands that reliably distinguish the objects?

Glossary

Accuracy
Freedom from error, lack of bias, close to true values. Note that accuracy is distinct from precision.

Active System
In remote sensing, a system that provides its own source of the electromagnetic radiation to be reflected or scattered by the object being sensed. For example, radar is an *active* microwave system since the radar system emits energy which is later detected. In contrast to a *passive* system.

Algorithm
A procedure for performing a specific action. In computer software, an algorithm is a set of instructions to the computer.

Alphanumeric
Consisting of both letters and numbers.

Analog
The representation of continuous numerical quantities (voltage, current, brightness, etc.) as opposed to discrete or "digital" units.

Ancillary Data
Additional, supplemental data.

Angular Field of View
See *Field of View* (FOV).

Array Processor
A specialized computing device used with a general-purpose computer. An

array processor performs mathematical operations on rectangular arrays of data at great speed. Frequently used in image processing and signal analysis.

Automated Geographic Information System (AGIS)
A geographic information system based on digital computers.

Band
A range of wavelengths of electromagnetic radiation; synonym for the term *channel*.

Band Interleaved by Line (BIL)
A specific implementation of a multivariate raster dataset. For each line in the raster, the values of each of the variables or bands are stored in sequence, before the set for the succeeding line.

Band Interleaved by Pixel (BIP)
A specific implementation of a multivariate raster dataset. For each pixel, the values of each of the variables or bands are stored before moving on to the next pixel.

Band Sequential (BSQ)
A specific implementation of a multivariate raster dataset. The complete data array for each separate variable or band is stored independently of the other variables.

Batch Processing
A computer processing procedure that is run at one time, without operator intervention, in contrast to *interactive processing*.

Bit
In digital computing, a binary digit, the smallest element of information. Two states are possible: "one" and "zero", also called "on" and "off", or "true"and "false."

Bits Per Inch (BPI)
A measurement of linear data density on magnetic storage materials such as tape.

Bits Per Second (BPS)

A measurement of the speed of data communications.

Byte

In a digital computer, a collection of 8 bits, and thus, 2^8 or 256 unique values. One byte can represent any of 256 discrete values; often as an integer between 0 and 255, or in a different system, any integer between -127 and +128.

Calibration

Determining the relationship between measurements with an instrument and standard or "known" values.

Camera

A device for recording images on a sensitized substrate, using one or more lenses to focus the light on the substrate, and a shutter to delimit the exposure.

Cartography

The science and practice of representing the features of the Earth's surface graphically.

Catalog

A collection of detailed information about whole datasets.

Cathode Ray Tube (CRT)

A device for displaying data electronically. The display surface is covered by a material that emits light when it is struck by an electron beam, which is steered by electrical signals. Used as the display device in direct-view televisions and many computer displays.

Census Tract

In the U.S., small areas averaging 4000 in population. They represent neighborhoods having similar socioeconomic characteristics and are defined in cooperation with the U.S. Bureau of the Census.

Central Processing Unit (CPU)

The part of a digital computer, that is responsible for numerical calculations and control.

Channel
In remote sensing, a specified spectral interval; also referred to as a *band*.

Coastal Zone Color Scanner (CZCS)
A multispectral sensor that flew on the Nimbus 7 satellite. The sensor was designed to register the color of ocean water as an indication of the amount of phytoplankton and other matter in the water.

Color Infrared Film (CIR)
A film emulsion with sensitivity from green through infrared. This emulsion is in common use, in part due to its sensitivity to different characteristics of live vegetation.

Computer Compatible Tape (CCT)
Magnetic tape containing data in computer-readable digital format. The most common format is termed *9-track*, for the number of parallel data recording tracks; frequently used linear data densities on the tracks are 1600 and 6250 bits per inch.

Contour Map
A topographic map that portrays relief by the use of lines indicating equal elevation. Such iso-elevation lines are termed *contour lines*.

Contrast
The differences in intensity (bright versus dark) of different parts of an image. The highest possible contrast is between black and white.

Contrast Enhancement
A procedure for artificially increasing the contrast in an image, often used to improve an analyst's ability to discern features.

Coordinate
A set of numerical quantities that designate position in a given reference system.

Dataset
A collection of similarly formatted records having like information (from one or more data sources).

Dataset Granule
A segment of a dataset, such as a single image, or one or several records in some fixed time increment, or files on each of several tapes. Usually considered the smallest element of information of value to a user.

Data System
A composite of capabilities for managing data, combining functions for maintaining a data directory, catalog, and inventory, as well as accessing data.

Defense Mapping Agency (DMA)
An agency of the U.S. Department of Defense.

Digital Computer
An electronic calculating device that operates on discrete variable representations, rather than with continuous or analog values (compare with "analog computer").

Digital Elevation Model (DEM)
A raster array of elevation values.

Digital Line Graph (DLG)
A vector data type produced by the U.S. Geological Survey.

Digital Number (DN)
In remote sensing, the numerical value of a specific pixel.

Digital to Analog Converter (D/A)
A device used to convert digital computer values to analog voltages. Found in computer display systems, where the digital computer values are converted to analog signals to manipulate the electron beam in a video display.

Digitizer
A device that converts analog information into a digital form. For flat graphic material, a digitizer can be either *flatbed* or *scanning*. For analog signals, a

digitizer consists of a digital-to-analog converter plus associated control and interface devices.

Directory
A collection of high-level information about whole datasets, for example, a directory could consist of dataset name, location where it is stored, and sources of further information.

Disks
A kind of direct-access storage device. Data is stored as coded areas on the face of one or more platters. With magnetic materials, flux changes are used to code for bits; with optical materials, changes in the refraction or reflection of light are used. Platters may have a data surface on one or both sides. The concentric circles of data on a platter are called tracks. These platters rotate rapidly and each bit of data is available to the reading/writing head once per revolution, providing high-speed random data access.

Dual Independent Map Encoding (DIME)
A form of vector data structure.

Easement
A restriction on land use. For example, an easement may prohibit a landowner from construction that would prevent emergency vehicles from gaining access to the property.

Electromagnetic Radiation
Energy propagating through space or matter in the form of waves of interacting electric and magnetic fields. Radio waves and light are examples of electromagnetic radiation.

Electromagnetic Spectrum
The range of wavelengths of known electromagnetic radiations, including ultraviolet radiation, visible radiation, infrared radiation, and microwaves.

End of File (EOF)
A specific computer code used to indicate the end of a dataset.

False Color Composite
An image formed by assigning colors arbitrarily to two or more black-and-white images of a single scene. The resulting image pinpoints differences and similarities between the original images; the input data frequently comes from the different wavelength bands of a multispectral sensor.

Field of View (FOV)
The solid angle through which an instrument is sensitive to radiation.

Footprint
The area on the surface being investigated by a remote sensing device; it is approximately given by the product of the beam width (in radians) times the altitude of the remote sensing platform.

Format
The physical organization of data elements within a dataset.

Fortran
A commonly used high-level algebraic language for coding computer algorithms; derived from FORmula TRANslation.

Geographic Information System (GIS)
The complete sequence of components for acquiring, processing, storing, and managing spatial data.

Geostationary Satellite
A satellite orbiting above the earth's equator at an altitude of approximately 36,000 km, such that its period of revolution about the Earth matches the Earth's rotational period. Thus, the satellite continuously views the same portion of the Earth's surface. Also called *geosynchronous satellite.*

Ground Truth
Information obtained on the ground, at the same time a remote sensing system is acquiring data from the same location. Ground truth is normally considered the most accurate available, and is used to interpret and calibrate remotely sensed observations.

Hardware
The physical components of a computer: central processing unit, memory, disk storage, tape drives, etc.

High-Density Digital Tape (HDDT)
An analog system for storing very high rate data that is not directly compatible with general-purpose computer systems.

High-Level Programming Language
A programming language (such as Fortran or Pascal) that is relatively close to written language or mathematics, and thus, relatively distant from the machine instructions executed by a computer.

Hotine Oblique Mercator (HOM)
A map projection, often used for data from spacecraft platforms.

Human Engineering
The design and implementation of automated and mechanical systems, taking human psychological concerns into consideration.

Image
A two-dimensional data representation. Examples include a photograph, a multispectral imaging sensor's data output, and the processed result of an aeromagnetic survey.

Image Classification
Analysis of digital image values, including one or more spatial, temporal, and spectral band relationships, to obtain categories of information about specific features.

Information
Data that has been processed for a particular use.

Instantaneous Field of View (IFOV)
For a sensing device, the area covered at a single moment, described either as the angle through which the sensor gathers radiation, or the area on the ground at a specified altitude.

Integrated Geographic Information System (IGIS)
A geographic information system that includes facilities for working with remotely sensed data.

Interactive Processing
Operation of a computer system through continual, instantaneous communication between man and machine. In contrast to *batch processing.*

Interval
(1) The time between two events. (2) As a measurement scale, variables in which the distances between data values are meaningful but based on an arbitrary starting point (e.g., Celsius and Fahrenheit temperature scales).

Inter-record Gap (IRG)
The effectively empty distance between data records on magnetic tape storage.

Inventory
A collection of information about the granules of a dataset.

Land Cover
The materials covering a land surface, such as vegetation, exposed soils, and so forth.

Landsat
A series of Earth-observing satellites (originally named ERTS, for Earth Resources Technology Satellite), first launched in 1972 by NASA, that serve as platforms for several instruments, including the return beam vidicon, the multispectral scanner, and the thematic mapper.

Land Use
Human-imposed functions of a land area.

Latitude
A system for indicating location north or south of the equator, based on the minimum angle subtended between the location in question, the Earth's center, and the equator.

LIDAR
The abbreviation for *light detection and ranging*. A remote sensing system, using the interaction of light pulses with materials at a distance. Used for determining atmospheric constituents and range of distant surfaces, among other things.

Longitude
An arbitrary system for indicating location east or west, based on the angular distance around the Earth's surface from the half of a great circle passing between the north and south poles and through Greenwich, England.

Low-Level Programming Language
A programming language very close to a computer's actual instruction set, working with the essential electronic elements of the computer; in contrast to high-level language.

Magnetometer
A device for measuring magnetic fields.

Manual Classification
The manual identification of features on aerial or satellite photographs by tone, color, texture, pattern, shape, and size.

Map
Usually a two-dimensional representation of all or part of the Earth's surface, showing selected natural or man-made features or data, preferably constructed on a definite projection with a specified scale.

Maximum Likelihood Rule
A decision criterion for assisting in the assignment of features to different classes. Where there is confusion between classes, assignments are made to the class of the highest probability.

Measurement Scale
A system for quantifying observations according to predetermined rules, which define four successfully greater levels of data precision (nominal, ordinal, interval, and ratio).

Medium

A physical device on which data is stored, such as a photographic product, map, or magnetic tape.

Multispectral Scanner (MSS)

Generally, a remote sensing device that records electromagnetic energy in several wavelength bands simultaneously. Specifically, a four-channel sensor carried by the Landsat series of satellites, first launched by NASA in 1972, with approximately an 80-meter IFOV.

Nadir

The point on the ground directly beneath the center of the remote sensing system.

National Science Foundation (NSF)

An independent agency of the U.S. federal government.

Nimbus

A series of Earth-observing experimental weather satellites carrying a variety of sensors. The Nimbus program was operated by the U.S. National Oceanic and Atmospheric Administration. The last of the series (Nimbus-7) was launched late in 1978.

Nominal

As a measurement scale, distinguishes things in terms of discrete categories, such as urban versus rural.

Oblique or Off-Nadir Viewing

A remote sensing system in which the observed electromagnetic radiation does not come from the vertical or nadir direction.

Operating System (OS)

The high-level administrative program running in a computer at all times. This program controls the overall operation of the computer and its tasks. Operating systems are usually hardware-specific, with the notable exception of the UNIX operating system.

Ordinal
As a measurement scale, distinguishes categories on the basis of rank by some quantitative measure (e.g., small, medium, large).

Orthophoto
A photograph that has been manipulated in such a way as to eliminate image displacement due to photographic tilt and relief.

Passive System
In remote sensing, a system that detects electromagnetic radiation that either has been emitted by the target object or has come from a natural source, such as the sun, and has been reflected or scattered by the target object. In contrast to an *active system*.

Photogrammetry
The techniques of obtaining precise measurements from images.

Pixel
Of a surface: the smallest unit whose characteristics may be uniquely determined. From *pic*ture *el*ement.

Planimetric Map
A map designed to portray the horizontal positions of features; vertical information is specifically ignored.

Platform
In remote sensing, the physical object (e.g., balloon, rocket, or satellite) that carries the instrument or sensor that makes the remote measurement.

Plotter
A device used to record information, such as graphs or maps, on paper or film. Frequently based on one or more pens that are moved over the medium under control of a computer.

Precision
The degrees of exactness with which a quantity is stated; this is directly related to the number of significant figures used in a description. A measurement that

divides phenomena into 10 intervals has less precision than one that divides the same phenomena into 100 intervals.

Preprocessing
The manipulation of data to make it suitable for further analysis. Often includes geometric modifications to bring several sets of data into a common map projection and scale, as well as scaling and reclassification.

Principal Point
On an aerial photograph, the point that corresponds to the axis of the lens system; commonly taken as the exact center of the photograph.

Projection
A systematic construction of features on a plane surface to represent corresponding features on a spherical surface. These features include observable phenomena (e.g., physical and cultural features such as coastlines and highways) as well as constructs (e.g., lines to represent parallels and meridians, political boundaries, or statistical units).

Quality Assurance (QA)
Referring to procedures for ensuring that an activity meets specified accuracy and precision goals.

Radiance
The flux of radiant (electromagnetic) energy measured in power units (e.g., watts).

Radiometer
A passive device for intercepting and quantitatively measuring electromagnetic radiation in a band of wavelengths.

Ratio
As a measurement scale, distinguishes things on the basis of magnitudes, such as length or age, where the multiplication and division of values have meaning, since the starting location (or zero-point) is not arbitrary.

Rectification
The process of projecting data onto a plane to bring the geometric

characteristics of data into conformance with a known coordinate system. For example, an oblique photograph could be rectified so that the location of features on the photograph correspond with their locations on a planimetric map.

Registration
Superposition of locations on one image with the corresponding locations on a second image or map.

Remote Sensing
Obtaining information about an object or phenomenon without direct contact.

Resampling
A set of mathematical procedures for changing the geometric characteristics of spatial data. Used in the processes of rectification and registration.

Resolution
(1) An indication of the number of component parts or units in a measurement system. When considering spatially referenced data, sometimes the number of units is normalized per unit area. (2) The minimum size of a feature that can be reliably distinguished by a remote sensing system.

Scalar
A quantity with a numeric value.

Scale
The ratio of distance on a map, chart, or image to the equivalent distance on the Earth's surface.

Seasat
An Earth-observing satellite designed to gather information about the oceans. Seasat was launched by NASA in 1978, but died abruptly later in that year. Data from Seasat's synthetic aperture radar is still in common use.

Sensor
A device that gathers electromagnetic radiation or other physical data and presents it in a form suitable for obtaining information about the environment.

Signal-to-Noise Ratio (S/N or SNR)
A determination that provides information about the quality of a measured value.

Signature
A set of spectral, tonal, temporal, or spatial characteristics that together serve to identify a class or feature by remote sensing.

Software
A computer program as written in a high- or low-level language.

Spatial Data
Data or information with implicit or explicit information about location.

Spatial Data Integration
The process of combining multiple spatial datasets and providing for their storage, retrieval, analysis, and display.

Spectrometer
A radiometer with a dispersive element (prism, grating, or circular interference filter) that enables characteristics of the incident radiation to be determined as a function of wavelength.

SPOT
Système Probatoire d'Observation de la Terre. A multispectral remote sensing satellite system with pointable sensors. The IFOV of the system is 20 meters in multispectral mode, and 10 meters in monochromatic mode. This French system was first launched in 1986.

Supervised Classification
Classification of data into discrete categories based on statistics developed from training sites.

Swath Width
The area on either side of a platform that is surveyed by a remote sensing instrument.

Synthetic Aperture Radar (SAR)
A radar system based on a series of elemental antenna units or sequences of observations from a single antenna, from which the effective antenna is mathematically constructed through signal processing. Such a system permits scanning of the signal through space without the need for moving antenna parts.

Tag
A descriptive element in a database. A database field such as "UPDATED=01/10/88" is composed of a tag to indicate the meaning of the data that follow, and the data itself (in this case, a date).

Thematic Mapper (TM)
A seven-channel, 30-meter IFOV multispectral scanner, designed for monitoring Earth resources. Flown by NASA on the Landsat 4 and 5 platforms.

Topography
The collective features of the surface of the Earth, including relief, hydrography, and cultural features.

Training Sites
Recognizable areas on an image with distinct spectral reflectance or other properties useful for identifying other similar areas.

Tuple
A data element composed of more than one value. For example, geographic locations are often specified by two values: latitude and longitude. The pair of latitude-longitude is called a tuple.

Ultraviolet (UV)
Electromagnetic radiation with wavelengths of 1.5 to 400 nanometers.

Unsupervised Classification
Automated classification of data into discrete categories by grouping together similar observations without the aid of training data. Also called *statistical clustering*.

Universal Transverse Mercator (UTM)
A common system for locations on the Earth's surface, based upon ground distances. A series of north-south zones are established, and locations are designated in terms of distance in meters east of the western edge of the zone, and north (or south) of the equator.

Variable
An unknown quantity. There are four essential kinds of variables: nominal, ordinal, interval, and ratio.

Vector
Generally, a quantity possessing both numerical value and direction. In terms of GIS, typically representing a boundary between spatial objects.

Video Information Communication and Retrieval (VICAR)
An iamge processing and raster GIS system developed by NASA's Jet Propulsion Laboratory.

Zoom Transfer Scope (ZTS)
An optical device for superimposing photographs and graphics on top of maps or other graphics, often used to update maps rapidly by direct tracing.

Abbreviations

AGIS
Automated Geographic Information System

AID
Agency for International Development

BIL
Band Interleaved by Line

BIP
Band Interleaved by Pixel

BPI
Bits Per Inch

BPS
Bits Per Second

BSQ
Band Sequential

B/W
Black and White

CBD
Central Business District

CCRS
Canada Centre for Remote Sensing

CCT
Computer Compatible Tape

CIR
Color Infrared Film

CPU
Central Processing Unit

CRT
Cathode Ray Tube

CZCS
Coastal Zone Color Scanner

D/A
Digital to Analog Converter

DEM
Digital Elevation Model

DIME
Dual Independent Map Encoding

DLG
Digital Line Graph

DMA
Defense Mapping Agency

DN
Digital Number

EOF
End of File

EPA
Environmental Protection Agency

ESA
European Space Agency

FL
Focal Length

FOV
Field of View

GBIS
Geo-Based Information System

HDDT
High-Density Digital Tape

HOM
Hotine Oblique Mercator

IFOV
Instantaneous Field of View

IGIS
Integrated Geographic Information System

IRG
Inter-record Gap

MSS
Multispectral Scanner

NASA
National Aeronautics and Space Administration

NOAA
National Oceanic and Atmospheric Administration

NSF
National Science Foundation

OS
Operating System

PI
Photo Interpretation or Photo Interpreter

QA
Quality Assurance

SAR
Synthetic Aperture Radar

SCA
Suitability/Capability Analysis

SNR or S/N
Signal to Noise Ratio

SPOT
Système Probatoire d'Observation de la Terre

TM
Thematic Mapper

UNESCO
United Nations Educational, Scientific, and Cultural Organization

UTM
Universal Transverse Mercator

UV
Ultraviolet

VICAR
Video Information Communication and Retrieval

ZTS
Zoom Transfer Scope

Bibliography

Allder, W.R., and A.A. Elasal, 1984. *Digital Line Graphs from 1:24,000-Scale Maps*. Circular 895-C, U.S. Geological Survey, Reston, Virginia.

Allder, W.R., A.J. Sziede, R.B. McEwen, and F.J. Beck, 1984. *USGS Digital Cartographic Data Standards: Digital Line Graph Attribute Coding Standards*. Circular 895-G, U.S. Geological Survey, Reston, Virginia.

Amidon, E.L., 1964. *A Computer-Oriented System for Assembling and Displaying Land Management Information*. U.S. Forest Service Research Paper PSW-17, Berkeley, California.

Anderson, J.M. and E.M. Mikhail, 1985. *Introduction to Surveying*. New York: McGraw-Hill Book Co.

Brooks, F.P. Jr., 1975. *The Mythical Man-Month: Essays on Software Engineering*. Reading, Massachusetts: Addison-Wesley Publishing Co.

Burgess, T.M. and R. Webster, 1980. "Optimal Interpolation and Isarithmic Mapping of Soil Properties: I. The Semi-Variogram and Punctual Kriging." *Journal of Soil Science*, Vol. 31, pp. 315-331.

Burrough, P.A., 1986. *Principles of Geographical Information Systems for Earth Resources Assessment*. Oxford: Clarendon Press.

Burt, P.J., 1980. "Tree and pyramid structures for coding Hexagonally Sampled Binary Images." *Computer Graphics and Image Processing*, Vol. 14, No. 3, pp. 271-280.

Calkins, H.W. and R.F. Tomlinson, 1977. *Geographic Information Systems: Methods and Equipment for Land Use Planning*. International Geographic Union Commission on Geographical Data Sensing and Processing. Resource and Land Investigations (RALI) Program, U.S. Geological Survey, Reston, Virginia.

Carson, R.L., 1962. *The Silent Spring*. Boston: Houghton-Mifflin Co.

Carter, J.R., 1984. *Computer Mapping: Progress in the 80's*. Washington, D.C.: Association of American Geographers.

Chang, Ning-San, 1981. *Image Analysis and Image Database Management*. Ann Arbor, Michigan: UMI Research Press.

Chin, R.T. and C.R. Dyer, 1986. "Model-based recognition in robot vision." *Computing Surveys*, Vol. 18, No. 1, pp. 67-108.

Cicone, R.C., 1977. Remote Sensing and Geographically Based Information Systems, *Proceedings of the 11th International Symposium on Remote Sensing of the Environment*, Vol. 2, p. 1130.

Colby, C.C., 1936. "Currents of Geographical Thought." *Annals AAG*, Vol. 26, pp. 1-36.

Colwell, R.N., ed., 1983. *The Manual of Remote Sensing*. Falls Church, Virginia: American Society of Photogrammetry.

Congalton, R.G., 1988. "A Comparison of Sampling Schemes Used in Generating Error Matrices for Assessing the Accuracy of MAps Generated from Remotely Sensed Data." *Photogrammetric Engineering and Remote Sensing*, Vol. 54, No. 5, pp. 593-600.

Cooke, D.F. and W.F. Maxfield, 1967. The Development of a Geographic Base File and its Uses for Mapping, *Urban and Regional Information System for Social Programs*, Proceedings of the 5th Annual Conference of the Urban and Regional Information Systems Association. pp. 207-218.

Cooke, D.F., 1987. "Map Storage on CD-ROM." *Byte Magazine*, Vol. 12, No.

8, pp. 129-138.

Coulson, R.N., L.J. Folse, and D.K. Loh, 1987. "Artificial Intelligence and Natural Resource Management." *Science*, Vol. 237, pp. 262-267.

Cowen, D.J., 1987. *GIS vs. CAD vs. DBMS: What are the Differences?* San Francisco: GIS '87, pp. 46-56.

Crapper, P.F., 1984. "An Estimate of the Number of Boundary Cells in a Mapped Landscape Coded to Grid Cells". *Photogrammetric Engineering and Remote Sensing*, Vol. 50, No. 10, pp. 1497-1503.

Dangermond, J., 1986. "CAD vs. GIS." *Computer Graphics World*, Vol. 9, No. 10, pp. 73-74.

Deuker, K.J., 1979. Land Resource Information Systems: Spatial and Attribute Resolution Issues, *Proceedings, Int. Symposium on Cartography and Computing: Auto-Carto IV*, Vol. 2, pp. 328-336.

Domaratz, M.A., C.A. Hallam, W.E. Schmidt and H.W. Calkins. 1983. *USGS Digital Cartographic Data Standards: Digital Line Graphs from 1:2,000,000-Scale Maps*. Circular 895-D, U.S. Geological Survey, Reston, Virginia.

Douglas, D.H. and A.R. Boyle, eds., 1982. *Computer Assisted Cartography and Information Processing: Hope and Realism*. Ottawa, Canada: Canadian Cartographic Association.

Doyle, F.J., 1985. "The Large Format Camera on Shuttle Mission 41-G." *Photogrammetric Engineering and Remote Sensing*, Vol. 51, No. 2, pp. 200-203.

Durfee, R.C., 1974. *ORRMIS: Oak Ridge Regional Modeling Information System*. Oak Ridge, Tennessee: Oak Ridge National Laboratory, p. 19.

Elassal, A.A. and V.M. Caruso. 1983. *USGS Digital Cartographic Data Standards: Digital Elevation Models*. Circular 895-B, U.S. Geological Survey, Reston, Virginia.

Enderle, G., K. Kansy and G. Pfaff, 1984. *Computer Graphics Programming:*

GKS-The Graphics Standard. New York: Springer-Verlag.

Estes, J.E., 1984. Improved Information Systems: A Critical Need. *Proceedings 10th Int. Symp. Machine Processing of Remotely Sensed Data.* Purdue University Laboratory for Applications of Remote Sensing, pp. 2-8.

Estes, J.E., J. Scepan, L. Ritter, and H.M. Borella, 1980. *Evaluation of Low-Altitude Remote Sensing Techniques for Obtaining Site Information.* Washington, D.C.: Nuclear Regulatory Commission, NUREG/CR-3583, S-762-RE.

Fegeas, R.G., R.W. Claire, S.C. Guptill, K.E. Anderson and C.A. Hallam. 1983. *USGS Digital Cartographic Data Standards: Land Use and Land Cover Digital Data*. Circular 895-E, U.S. Geological Survey, Reston, Virginia.

Foley, J.D. and A. Van Dam, 1982. *Fundamentals of Interactive Computer Graphics*. Reading, Massachusetts: Addison-Wesley Publishing Co.

Ford, G.E. and C.I. Zanelli, 1985. "Analysis and Quantification of Errors in the Geometric Correction of Satellite Images." *Photogrammetric Engineering and Remote Sensing*, Vol. 51, No. 11, pp. 1725-1734.

Foresman, T.W., 1986. "Mapping, Monitoring and Modeling of Hazardous Waste Sites." *The Science of the Total Environment*. Vol. 56, pp. 255-264.

Freeman, H. and J. Ahn, 1984. Autonap: an Expert System for Automatic Map Name Placement. *Proc. International Symposium on Spatial Data Handling*. Zurich. Vol. 2, pp. 544-569.

Gaits, G.M., 1969. "Thematic Mapping by Computer." *Cartographic Journal*. Vol. 6, No. 1, pp. 50-68.

Gibson, L. and D. Lucas, 1982. Vector of Raster Images using Hierarchical Methods. *Computer Graphics and Image Processing*, Vol. 20, pp. 82-89.

Goetz, A.F.H., B.N. Rock, and L.C. Rowan, 1983. "Remote Sensing for Exploration: An Overview." *Economic Geology*, Vol. 78, No. 4, pp. 573-684.

Goetz, A.F.H., G. Vane, J.E. Solomon, and B.N. Rock, 1985. "Imaging Spectrometry for Earth Remote Sensing." *Science*, Vol. 228, No. 4704, pp. 1147-1153.

Goodchild, M.F., 1980. "Fractals and the Accuracy of Geographic Measures." *Mathematical Geology*. Vol. 12, No. 2, pp. 85-98.

Goodchild, M.F., 1987. "A Spatial Analytical Perspective on Geographical Information Systems." *Int. J. Geographical Information Systems*, Vol. 1, No. 4, pp. 327-334.

Goodchild, M.F. and B.R. Rizzo, 1987. "Performance Evaluation and Work-load Estimation for Geographic Information Systems". *Int. J. Geographical Information Systems*, Vol. 1, No. 1, pp. 67-76.

Green, R., 1964. *The Storage and Retrieval of Data for Water Quality Control*. Public Health Service Publication No. 1263, U.S. Department of Health, Education, and Welfare, Public Health Service. Washington, D.C.

Hansen, H., 1987. *Justification for a Management Information System*. San Francisco: GIS '87, pp. 19-28.

Hardisky, M.A., R.M. Smart, and V. Klemas, 1983. "Seasonal Spectral Characteristics and Aboveground Biomass of the Tidal Marsh Plant Spartina alterniflora. *Photogrammetric Engineering and Remote Sensing*, Vol. 49, No. 1, pp. 85-92.

Harley, J.B., B.B. Petchenik, and L.W. Towner, 1978. *Mapping the American Revolutionary War*. Urbana, Illinois: University of Illinois Press.

Hewitt, M.J. III and E.N. Koglin, 1987. *A Planning Strategy for using GIS in the assessment of Environmental Problems: A Customer's Guide*. San Francisco: GIS '87, pp. 128-147.

Jarvenpaa, S.L. and G.W. Dickson, 1988. "Graphics and Managerial Decision Making: Research Based Guidelines." *Communications of the ACM*, Vol. 31, No. 6, pp. 764-774.

Jensen, J.R. and E.J. Christensen, 1986. "Solid and Hazardous Waste Disposal Site Selection using Digital Geographic Information System Techniques." *The Science of the Total Environment*, Vol. 56, pp. 265-276.

Jensen, J.R., E.W. Ramsey, H.E. Mackey Jr., E.J. Christensen, and R. R. Sharitz, 1987. "Inland Wetland Change Detection using Aircraft MSS Data." *Photogrammetric Engineering and Remote Sensing*, Vol. 53. No. 5, pp. 521-529.

Kemper, A. and M. Wallrath, 1987. "An Analysis of Geometric Modeling in Database Systems." *ACM Computing Surveys*, Vol. 19, No. 1, pp. 47-91.

Kennedy, M. and C. Guinn, 1975. *Automated Spatial Data Information Systems: Avoiding Failure*. Louisville, Kentucky: Urban Studies Center.

Kennedy, M. and C.R. Meyers, 1977. *Spatial Information Systems: Introduction*, Louisville, Kentucky: Urban Studies Center.

Knapp, E., 1978. *Landsat and Ancillary Data Inputs to an Automated Geographic Information System*, Report No. CSC/tr-78/6019. Silver Springs, MD: Computer Science Corp.

Kok, A.L., J.A.R. Blais and R.M. Rangayyan, 1987. "Filtering of Digitally Corrected Gestalt Elevation Data." *Photogrammetric Engineering and Remote Sensing*, Vol. 53, No. 5, pp. 535-538.

LeDrew, E.F. and S.E. Franklin, 1985. "The Use of Thermal Infrared Imagery in Surface Current Analysis of a Small Lake." *Photogrammetric Engineering and Remote Sensing*, Vol. 51, No. 5, pp. 565-573.

Lemmon, H., 1986. "Comax: An Expert System for Cotton Crop Management." *Science*, Vol. 233, pp. 29-33.

Lewis, J., 1985. "The Birth of EPA." *EPA Journal*, Vol. II, No. 9, United States Environmental Protection Agency, pp. 6-11.

Light, D.L., 1983. "Mass Storage Estimates for the Digital Mapping Era." *Technical Papers of the 43rd Annual Meeting of the American Congress on Surveying and Mapping*. Washington, D.C., March 1983.

Lillesand, T.M. and R.W. Kiefer, 1987. *Remote Sensing and Image Interpretation*, 2nd edition. New York: John Wiley and Sons.

Lorie, R.A. and A. Meier, 1984. "Using a Relational Database for Geographical Databases." *Geo-Processing*, Vol. 2, pp. 243-257.

MacDonald, R.B. and F.G. Hall, 1981, "Global Crop Forecasting," *Science*, Vol. 208, pp. 670-679.

MacDonald, R.B., A.G. Houston, R.P. Heydorn, D.B. Botkin, J.E. Estes and A. H. Strahler, 1981. *Monitoring Global Vegetation*, Reprint of Paper for 7th International Symposium on Machine Processing of Remotely Sensed Data. Houston, Texas: Earth Resources Sensing Division, Space and Life Science Directorate, NASA.

Males, R.M., 1977. *ADAPT-A Spatial Data Structure for Use with Planning and Design Models*, Working Papers from the Advanced Study Symposium on Topological Data Structures for Geographic Information Systems, Vol. 2, pp. 1-35.

Marble, D.F. and D.J. Peuquet, 1977. *Computer Software for Spatial Data Handling: Current Status and Future Development Needs*. Buffalo, New York: Geographic Information Systems Laboratory, State University of New York.

Marble, D.F., H.W. Calkins, and D.J. Peuquet, 1984. *Basic Readings in Geographic Information Systems*. Williamsville, New York: Spad Systems, Ltd.

Mark, D.M. and P.B. Aronson, 1984. "Scale-Dependent Fractal Dimensions of Topographic Surfaces: An Empirical Investigation, with Applications in Geomorphology and Computer Mapping." *Mathematical Geology*, Vol. 16, No. 7, pp. 671-683.

Martin, J. 1975. *Computer Data-Base Organization*. Englewood Cliffs, New Jersey: Prentice-Hall, Inc.

Martin, D. 1986. *Advanced Database Techniques*. Cambridge, Massachusetts: MIT Press.

McEwen, R.B., H.W. Calkins, and B.S. Ramey, 1983. *USGS Digital Cartographic Data Standards: Overview and USGS Activities*. Circular 895-A. U.S. Geological Survey, Reston, Virginia.

McHarg, I., 1969. *Design With Nature*. Garden City, New Jersey: Doubleday and Co.

Meynen, E., 1973. *Multilingual Dictionary of Technical Terms in Cartography*. International Cartographic Association, Commission II. Wiesbaden, West Germany: Franz Steiner Verlag GBMH.

Moik, J.G., 1980. *Digital Processing of Remotely Sensed Images*. Houston, Texas: National Aeronautics and Space Administration NASA SP431.

Monmonier, M. and G.A. Schnell, 1988. *Map Appreciation*. Englewood Cliffs, New Jersey: Prentice Hall.

Nagy, G. and S. Wagle, 1979. "Geographic Data Processing." *Computing Surveys*, Vol. 11, No. 2, pp. 139-181.

Naisbitt, J., 1984. *Megatrends*. New York: Warner Books, Inc.

Noyt-Meir, I. and R.H. Whittaker, 1977. "Continuous Multivariate Methods in Community Analysis: Some Problems and Developments." *Vegetatio*, Vol. 33, pp. 79-98.

Parent P. and R. Church, 1988. *Evolution of Geographic Information Systems as Decision Making Tools*. San Francisco: GIS '87. pp. 63-71.

Peucker, T.K. and N. Chrisman, 1975. "Cartographic Data Structures". *The American Cartographer*, Vol. 2, No. 1, pp. 55-69.

Peuquet, D.J., 1977. *Raster Data Handling in Geographic Information Systems*. Buffalo, New York: Geographic Information Systems Laboratory, State University of New York.

Peuquet, D.J., 1981a. "An Examination of Techniques for Reformatting Digital Cartographic Data. Part 1: The Raster-to-Vector Process."

Cartographica, Vol. 18, No. 1, pp. 34-48.

Peuquet, D.J., 1981b. "An Examination of Techniques for Reformatting Digital Cartographic Data. Part 2: The Vector-to-Raster Process." *Cartographica*, Vol. 18, No. 3, pp. 21-33.

Pountain, D., 1987. "Vector-To-Raster Algorithms." *Byte Magazine*, Sept. 1987, pp. 177-184.

Rice, H.C. Jr. and A.S.K. Brown (eds. and translators), 1972. *The American Campaigns of Rochambeau's Army*. Princeton, New Jersey: Princeton University Press.

Robinson, A., R. Sale and J. Morrison, 1978. *Elements of Cartography*. 4th ed. New York: John Wiley and Sons.

Robinson, B.F. and D.P. DeWitt, 1983. "Electro-Optical Non-Imaging Sensors." In *The Manual of Remote Sensing*. R.N. Colwell, ed. Falls Church, Virginia: American Society of Photogrammetry, pp. 293-333.

Rosenfeld, G.H., 1982. "Sample Design for Estimating Change in Land Use and Cover," *Photogrammetric Engineering and Remote Sensing*, Vol. 48, No.5, pp. 793-801.

Rosenfeld, G.H., K. Fitzpatrick-Lins, and H.S. Ling, 1981. "Sampling for Thematic Map Accuracy Testing." *Photogrammetric Engineering and Remote Sensing*, Vol. 48, No. 1, pp. 131-137.

Rowe, J.S., 1959. *Forest Regions of Canada*. Bulletin 123. Ottowa, Canada: Dept. Northern Affairs and Natural Resources, Forestry Branch.

Salmen, L., D.L. Mutter, and K. Burnham, 1977. *A General Design Schema for an Operational Geographic Information System for the U.S. Fish and Wildlife Service Region Six*. Fort Collins, Colorado: Western Governors Policy Office, p. 27.

Samet, H., 1985. "Data Structures for Quadtree Approximation and Compression." *Communications ACM*, Vol. 28, No. 9, pp. 973-993.

Samet, H., 1984. "The Quadtree and Relate Hierarchical Data Structures." *ACM Computing Surv.*, Vol. 16, No. 2, pp. 187-260.

Sauer, C.O., 1919. "Mapping Utilization of Land." *Geographical Review*. Vol. 8., pp. 47-54.

Scott, D.S. and S. S. Iyengar, 1986. "TID: A Translation Invariant Data Structure for Storing Images." *Communications ACM*, Vol. 29, No. 5, pp. 418-429.

Shelton, R.L. and J.E. Estes, 1979. "Integration of Remote Sensing and Geographic Information Systems," *Proceedings, 13th International Symposium on Remote Sensing of Environment.* Ann Arbor, Michigan: Environmental Research Institute of Michigan, pp. 675-692.

Short, N.M., 1982. *The Landsat Tutorial Workbook: Basics of Satellite Remote Sensing*. NASA Remote Sensing Pub. 1078. Washington, D.C.: U.S. Government Printing Office.

Simonett, D.S., ed., 1976. *Applications Review for a Space Program Imaging Radar*. Santa Barbara Remote Sensing Unit, Technical Report No. 1, NASA Contract No. NAS9-14816, Johnson Space Center, pp. 6-18.

Slater, P.N., 1983. "Photographic Systems for Remote Sensing." In *The Manual of Remote Sensing,* R.N. Colwell, ed. American Society of Photogrammetry, pp. 231-291.

Smith, B. and J. Wellington, 1986. *Initial Graphics Exchange Specifications (IGES), Version 3.0.* U.S. Department of Commerce NBSIR 86-3359.

Smith, T.R., 1984. *Knowledge-Based Approaches to Spatial Data Handling Systems-Background Materials to Workshop 3*, International Geographical Union, International Symposium on Spatial Data Handling. Zurich, Switzerland.

Smith, T.R., S. Menon, J.L. Star and J.E. Estes, 1987a. "Requirements and Principles for the implementation and construction of Large-scale Geographic

Information Systems." *Int. J. Geographical Information Systems*, Vol. 1, No. 1, pp. 13-32.

Smith, T.R., D.J. Peuquet, S. Menon and P. Agarwal, 1987b. "KBGIS II: A Knowledge Based Geographic Information System." *Int. J. Geographical Information Systems*, Vol. 1, No. 2, pp. 149-172.

Snyder, J.P., 1985. *Computer-Assisted Map Projection Research*. U.S. Department of the Interior, U.S. Geological Survey Bulletin 1629. Washington, D.C.: U.S. Government Printing Office, 157 pp.

Spanner, M.A., A.H. Strahler, and J.E. Estes, 1982. "Soil Loss Prediction in a Geographic Information System Format." *16th Int. Symposium on Remote Sensing of the Environment*, Buenos Aires, Argentina.

Star, J.L., M.J. Cosentino and T.W. Foresman, 1984. "Geographic Information Systems: Questions to Ask Before It's Too Late." *Proceedings 10th Int. Symp. on Machine Processing of Remotely Sensed Data*. Lafayette, Indiana: Purdue University Lab. for Applications of Remote Sensing, pp. 194-197.

Steinitz, C.F., P. Parker and L. Jordan, 1976. "Hand Drawn Overlays: Their History and Prospective Uses." *Landscape Architecture*, Vol. 66, pp. 444-455.

Streich, T.A., 1986. *Geographic Data Processing: A Contemporary Overview*. Master's thesis, University of California, Santa Barbara, Department of Geography.

System Development Corp., 1968. *Urban and Regional Information Systems: Support for Planning in Metropolitan Areas*. Washington, D.C.

Tailor, A., A. Cross, D.C. Hogg, and D.C. Mason, 1986. "Knowledge-based Interpretation of Remotely Sensed Images." *Image and Vision Computing*, Vol. 4, No. 2, pp. 67-83.

Thompson, M.M., ed., 1966. *Manual of Photogrammetry*, 3rd edition. Falls Church, Virginia: American Society of Photogrammetry.

Tinney, L.R. and C.E. Ezra, 1986. "Geographic Information Systems Applications for the United States Department of Energy's Comprehensive Integrated Remote Sensing Program." *Proc. Workshop on Geographic Information Systems for Environmental Protection*, Las Vegas, Nevada: Environmental Research Center, University of Nevada, Las Vegas, pp. 24-33.

Tobler, W.R., 1979. *Cellular Geography*. In S. Gale and G. Olsson, eds. *Philosophy in Geography*, Dordrecht, Holland: D. Riedel Publishing Company, pp. 379-386.

Tobler, W.R., 1987. "Measuring Spatial Resolution." *Proc. International Workshop on Geographic Information Systems*, in Chen, Shupeng, et al., eds. Beijing, pp. 42-48.

Tobler, W.R. and Z. Chen, 1986. "A Quadtree for Global Information Storage." *Geographical Analysis*, Vol. 18, No. 4, pp. 360-371.

Tomlinson, R.F., 1982. "Panel Discussion: Technology Alternatives and Technology Transfer." In *Computer Assisted Cartography and Geographic Information Processing, Hope, and Realism*, Douglas and Boyle, eds. Canadian Cartographic Association, Dept. of Geography, University of Ottawa, pp. 65-71.

Tucker, C.J., B.N. Holben and T.E. Goff, 1984. "Intensive Forest Clearing in Rondonia, Brazil as Detected by Satellite Remote Sensing." *Remote Sensing of Environment*, Vol. 15, pp. 255-261.

Tucker, C.J. and J.A. Gatlin, 1984. "Monitoring Vegetation in the Nile Delta with NOAA-6 and NOAA-7 AVHRR Imagery," *Photogrammetric Engineering and Remote Sensing*, Vol. 50, No. 1, pp. 53-61.

Tucker, C.J., J.R.G. Townshend and T.E. Goff, 1985. "African Land-Cover Classification Using Satellite Data." *Science*, Vol. 227, pp. 369-374.

Tufte, E.R., 1983. *The Visual Display of Quantitative Information*. Cheshire, Connecticut: Graphics Press.

Ullman, J.D., 1982. *Principles of Database Systems*, 2nd ed. Rockville,

Maryland: Computer Science Press.

U.S. Department of Commerce, Bureau of the Census, 1970. *The DIME Geocoding System*. Report #4, Census Use Study. Washington, D.C.

U.S. Department of the Interior, 1984. *Map Projections for use with the Geographic Information System*. Fish and Wildlife Service FWS/OBS-84/17, Washington, D.C.: U.S. Government Printing Office.

U.S. Fish and Wildlife Service, 1977. *Comparison of Selected Operational Capabilities of Fifty-four Geographic Information Systems*. FWS/OBS 77/54, Biological Services Program. Washington, D.C.: Government Printing Office.

Van Roessel, J.W., 1987. "Design of a Spatial Data Structure using Relational Normal Forms." *Int. J. Geographical Information Systems*, Vol. 1, No. 1, pp. 33-50.

Vlcek, J. and D. King, 1983. "Detection of Subsurface Soil Moisture by Thermal Sensing: Results of Laboratory, Close Range, and Aerial Studies." *Photogrammetric Engineering and Remote Sensing*, Vol. 49, No. 11, pp. 1593-1597.

Waldrop, M.M., 1984. "Natural Language Understanding." *Science*, Vol. 224, pp. 372-374.

Webster, R. and T.M. Burgess, 1980. "Optimal Interpolation and Isarithmic Mapping of Soil Properties: III. Changing Drift and Universal Kriging." *Journal of Soil Science*, Vol. 31, pp. 505-524.

Welch, R., T.R. Jordan, and M. Ehlers, 1985. "Comparative Evaluations of the Geodetic Accuracy and Cartographic Potential of Landsat-4 and Landsat-5 Thematic Mapper Image Data." *Photogrammetric Engineering and Remote Sensing*, Vol. 51, No. 9, pp. 1249-1262.

Winkler, R.L. and W.L. Hays, 1975. *Statistics: Probability, Inference and Decision*, 2nd ed. New York: Holt, Rinehart, and Winston.

Wolfe, P.R., 1983. *Elements of Photogrammetry*, 2nd ed. New York: McGraw-Hill Book Co.

Index

Aberrations, 200
Absolute, 29, 98, 112, 120, 124
Accuracy, 16, 22, 25, 28, 57, 62, 70, 85, 88, 96, 100, 123, 124, 151, 172, 191, 207, 210, 215, 222, 225, 235, 237, 245, 246
Acquisition, 24, 27, 61, 70, 78, 120, 126, 199, 205, 217, 225, 235, 242, 247
Across-track, 203, 204
Address, 19, 51, 61, 131, 201, 219
Adjacency, 90, 142
Aerial, 2, 6, 24, 43, 62, 66, 69, 75, 96, 98, 105, 111, 114, 118, 179, 191, 193, 197, 201, 209, 211, 221, 234
Aggregation, 68, 93, 114, 144, 146, 151, 152, 153
Aircraft, 77, 106, 118, 120, 124, 145, 156, 173, 192, 198, 201, 202, 231
Albers, 101, 102
Algorithms, 31, 41, 62, 82, 84, 96, 109, 143, 151, 161, 183, 210, 211, 215, 229, 248
Aliasing, 81, 186
Altitude, 63, 105, 199, 202, 205, 208, 232
Analog, 3, 17, 20, 66, 188, 193, 211, 241
Ancillary, 22, 33, 63, 82, 107, 119, 122, 191, 192, 205, 218, 227
Animation, 179
Anti-aliasing, 187
Aperture, 123, 209
Approximation, 104, 115, 235
Arc, 52, 80, 129, 147, 164, 234
Archive, 27, 50, 63, 69, 183, 224, 229
Arc-node, 49, 52, 59, 80, 94, 129, 136, 147
Arrays, 34, 47, 77, 107, 112, 137, 147, 199, 214
Aspect, 123, 163
Astigmatism, 200
Attribute, 17, 29, 33, 41, 45, 51, 80, 93, 107, 113, 132, 143, 151, 161, 164, 175, 184, 215, 227
Autocorrelation, 71
Averaging, 42, 93, 109, 114, 116, 152
Axis, 67, 84, 88, 90, 99, 154, 197, 200, 202
Azimuthal, 99, 101

Backup, 132, 140, 170, 223
Band, 77, 196, 201, 211
Bathymetry, 168, 226, 247
Benchmark, 30, 80, 143
Bias, 59, 70, 110, 251
BIL, 78, 79
Bilinear, 107, 108, 109
Binary, 41, 46, 81, 173, 214, 234
BIP, 78
Bit, 47, 130, 186, 204, 209, 215
Bit-per-pixel, 214
Black-and-white, 67, 186, 193, 209
Boolean, 148, 170, 234
Boundaries, 4, 7, 27, 32, 39, 47, 49, 55, 58, 62, 65, 68, 80, 83, 88, 91, 93, 98, 120, 141, 145, 156, 161, 164, 175, 177, 215, 218, 230, 236, 238, 240, 244
BPI, 224, 247
Bresenham's algorithm, 186
Brightness, 77, 88, 90, 107, 114, 119, 186, 207, 213, 232
Browse, 133
BSQ, 78
Buffer, 132, 158
Byte, 186, 204

Cache, 131
Calibrate, 71, 89, 101, 116, 119, 205, 226

Caliper, 124
Camera, 63, 67, 101, 105, 118, 124, 193, 202
Campaign, 75
Cardinal, 42
Cartesian, 34, 112
Cartogram, 179
Cartography, 15, 17, 19, 22
Cell, 33, 68, 80, 93, 107, 113, 147, 151, 156, 160, 167, 170, 178, 186, 202, 222, 224, 227
Census, 18, 50, 57, 65, 68, 72, 169, 175, 182
Centroid, 141, 156, 165
Chain, 2, 10, 40, 41, 49, 90
Chart, 180, 181
Choropleth, 175, 176, 177
Chromatic, 200
Circumference, 91, 98
Classification, 16, 62, 100, 149, 171, 192, 210, 217, 227, 237, 241
Clumped, 71
Cluster, 139, 151
Collateral, 122, 192
Color, 9, 30, 67, 88, 119, 121, 175, 183, 190, 193, 196, 210, 232
Coma, 200
Compression, 41, 213, 224
Concurrency, 137
Confidence, 150, 225
Conformal, 100, 103
Conic, 99
Containment, 90, 142, 244
Contextual, 210
Contiguous, 52, 230, 243
Continuous-tone, 209
Contour, 4, 27, 33, 35, 58, 67, 164, 176, 214, 231
Convolution, 108, 160
Coordinate, 7, 15, 34, 48, 52, 57, 64, 68, 80, 89, 98, 139, 154, 236
Correlation, 118, 125, 168
Corridor, 7, 157
Cost, 42, 54, 66, 82, 86, 123, 148, 172, 201, 238, 246
Covariance, 168
Covariate, 208
Coverage, 96
Cross-tabulation, 168, 169
Cubic interpolation, 108
Cursor, 86, 88
Curvature, 106, 154
Curvilinear, 182
Cylindrical, 100

Dasymetric, 177
Database, 2, 7, 26, 49, 68, 76, 91, 110, 115, 126, 157, 165, 181, 191, 222
Datum, 33, 240
Decision-making, 2, 61, 221
Deflection, 198
Deformation, 99
Demographic, 19, 26, 50, 61, 114, 168, 182, 226, 247
Densitometer, 88
Derivative, 162
Detection, 57, 76, 93, 122, 210, 246
Dichotomous key, 122
Digital Line Graph, 55
Digitize, 86, 95, 125, 191, 212, 236
DIME, 49, 50, 51, 52, 57, 58, 142
Dimensionality, 217
Discontinuities, 97, 120
Discriminate, 186, 213
Dispersed, 71, 200
Display, 15, 57, 97, 101, 156, 170, 210, 213, 221, 248
Distance, 11, 15, 28, 34, 66, 71, 75, 99, 100, 110, 115, 116, 124, 131, 150, 158, 164, 172, 177, 187, 197, 243
Distance-weighted, 118
Distortion, 6, 81, 96, 106, 154, 186, 199
Distributions, 63, 69, 74, 121, 161, 178, 207
Dynamic-range, 17

Easement, 4, 39, 157
Easting, 111
Edge-detection, 161, 211
Efficiencies, 224
Electrostatic, 86, 183, 188
Elevation, 4, 8, 11, 27, 32, 35, 42, 58, 62, 64, 68, 70, 77, 80, 88, 93, 114, 121, 133, 139, 143, 152, 162, 172, 177, 192, 197, 200, 217, 222, 232, 238, 242, 251
Emission, 206
Emulsion, 63, 195

Encoding, 41, 49, 57, 62, 90, 93, 133, 142
Endpoints, 51, 72
Energy, 65, 160, 196, 197, 204, 208, 209, 246
Enhance, 161, 207, 212
Equal-area, 101
Equivalence, 101
Estimation, 114, 240, 241
Exclusive-OR, 149
Expectations, 211
Exponential, 128
Exposure, 188, 197, 201
Extract, 62, 65, 79, 82, 96, 169, 181, 191, 211, 218
Extrapolate, 73, 96, 228

Far-infrared, 208
Fieldwork, 191
Film, 6, 27, 63, 77, 86, 106, 118, 183, 196
Filtering, 152, 160, 207, 236
Fine-grained, 120
Fixed-length, 126

Generalization, 17, 76, 91, 151, 244
Geocoding, 20, 97, 103
Geodetic, 7, 9, 33, 43, 88, 98, 103, 107, 113, 205
Geographers, 1, 18
Geography, 1, 19, 24, 119, 251
Geometry, 30, 33, 53, 67, 103, 119, 123, 149, 198, 202, 205, 217
Geoprocessing, 19, 22, 27, 247
Georeferencing, 34, 62, 76, 88, 96, 98, 104, 109
Geostationary, 12, 202
Gradient, 171, 177
Granules, 182
Gray-scale, 183

Half-tones, 67
Hardcopy, 229
Heuristics, 167, 183
Hexagon, 38
Hierarchical, 42, 52, 59, 95, 137, 142
Higher-resolution, 46
High-pass, 161, 236
Histogram, 167, 168, 181
Homogeneity, 77
Hypotheses, 24, 75, 122, 151, 168, 182, 206
Hypsography, 55, 68, 224

Illumination, 210, 216
Imagery, 62, 85, 118, 122, 154, 183, 186, 196, 201, 206, 210, 216, 231, 238
Image-processing, 207, 211, 214, 218, 241
Independence, 130, 140
Indexing, 141
Inductive, 243
Inferences, 70, 101, 121
Infrared, 121, 160, 193, 208, 214, 251
Integer, 81, 93, 128, 152, 178
Integration, 153, 191
Integrity, 129, 140, 224
Intensity, 119, 205
Interleaved, 78
Interpolaion, 76, 107, 114, 152, 156, 164, 177, 218, 226
Interpretation, 2, 4, 22, 37, 66, 119, 151, 192, 207, 210, 217, 238
Intersection, 52, 99, 147, 164
Interval, 28, 42, 74, 148, 160, 168, 177, 202, 216
Isarithmic, 177
Isolines, 172, 214

Kernel, 160
Keywords, 166
Knowledge-based, 211
Kriging, 117
Kurtosis, 167

Lambert, 101
Landcover, 156
Landform, 179
Landuse, 21
Large-format, 86
Latitude, 7, 29, 34, 63, 88, 98, 103, 106, 128, 207
Layers, 4, 7, 17, 20, 32, 42, 55, 68, 76, 143, 146, 154, 157, 168, 180, 196, 213, 222, 225, 230, 233

Least-cost, 172
Least-squares, 92
Length, 7, 10, 37, 53, 67, 93, 103, 111, 118, 124, 130, 159, 165, 171, 180, 190, 197, 202, 226, 229
Lens, 67, 106, 118, 193, 200, 203
Life-cycle, 249
Light, 88, 99, 121, 150, 188, 195, 200, 216, 223, 232
Location, 7, 15, 26, 29, 33, 45, 48, 50, 55, 62, 69, 78, 80, 86, 88, 93, 103, 106, 110, 113, 119, 139, 149, 156, 165, 171, 177, 188, 203, 214, 218, 222, 242, 249
Logical, 29, 63, 126, 129, 148, 205, 217
Longitude, 7, 29, 88, 98, 103, 106, 128, 202
Look-up, 178
Lower-resolution, 46, 93
Low-altitude, 200
Low-frequency, 160
Low-order, 116, 123
Low-pass, 160, 161

Machine-assisted, 191, 196, 209
Magnetic-tape, 132
Magnification, 197
Mainframe, 142
Majority, 37, 39, 66, 93, 152, 245
Manipulation, 24, 58, 76, 130, 141, 148, 161, 163, 169, 223, 247
Manuscript, 130, 223
Man-machine, 95
Man-Month, 229
Map, 2, 11, 15, 20, 27, 30, 49, 55, 62, 85, 95, 109, 123, 129, 137, 145, 154, 172, 184, 211, 215, 223, 231, 240, 245
Map-guided, 217
Matrix, 110, 146, 217
Maximize, 72, 114, 230, 248
Mean, 11, 33, 44, 70, 115, 142, 151, 160, 167, 202, 242
Measurement, 12, 28, 72, 100, 119, 122, 164, 192, 210, 226
Media, 77, 130, 139, 174
Medial, 84
Median, 121, 161, 167
Memory, 126, 131,
Mercator, 34, 68, 99, 110, 236
Merging, 59, 76, 95, 146, 151, 191, 232, 235
Meridian, 34, 103
Metadata, 128
Microfilm, 183
Microwave, 208
Middle-infrared, 208
Milestones, 228
Minicomputer, 142
Minimization, 130
Model, 12, 20, 24, 27, 32, 36, 54, 58, 65, 68, 99, 106, 114, 124, 134, 136, 165, 172, 222, 226, 231, 234, 238
Monitoring, 8, 12, 26, 74, 192, 215, 228, 241, 246
Monochrome, 186, 188
Mosaic, 96
Multispectral, 75, 77, 107, 145, 150, 172, 183, 191, 201, 203, 206, 210, 214, 227, 241
Multivariate, 77, 149, 215

Nadir, 199
Navigational, 137, 138
Nearest-neighbor, 107
Near-infrared, 208
Neighborhood, 36, 83, 93, 115, 157
Network, 8, 33, 55, 62, 133, 158, 171, 179, 230, 233, 236, 240, 247
Nodes, 9, 43, 47, 50, 80, 95, 107, 110, 113, 129, 147, 164, 242
Nominal, 28, 41, 93, 109, 114, 134, 145, 168, 202, 205, 218
Non-parametric, 161, 168
Non-procedural, 136
Northing, 111

Oblateness, 98
Oblique, 105, 179
Off-nadir, 106
Optimizing, 79, 130, 133
Orbit, 199, 202, 203
Ordinal, 28, 93, 109, 114, 145, 160, 163, 168
Orthographic, 106
Orthophoto, 67

Output, 13, 27, 44, 60, 77, 83, 101, 107, 123, 136, 147, 156, 169, 171, 174, 177, 183, 187, 190, 217, 235, 240, 247
Overlay, 6, 20, 37, 57, 98, 123, 146, 149, 191, 231, 236
Overpass, 203

Paging, 141
Panchromatic, 206, 224, 236
Parallax, 125
Parcel, 4, 150, 159, 165, 240
Partitioned, 230
Path, 30, 83, 92, 163, 170, 198
Patterns, 2, 4, 16, 71, 90, 99, 103, 121, 151, 161, 173, 175, 177, 179, 184, 203, 207, 217, 231
Perimeter, 140, 142, 164, 180, 243
Perspective, 6, 29, 101, 106, 152, 156, 182, 199, 209, 232, 248, 251
Phosphor, 190
Photodetector, 90, 198
Photogrammetry, 19, 66, 91, 101, 119, 195
Photograph, 4, 6, 11, 67, 75, 85, 96, 105, 114, 119, 120, 124, 199
Photointerpretation, 62, 76, 118, 122, 124, 214, 234
Physiography, 114, 226, 227
Pigments, 208
Pixel, 41, 46, 78, 80, 90, 109, 150, 154, 160, 164, 172, 184, 186, 203, 207, 213, 216, 227, 236, 243
Pixel-interleaved, 78
Planar, 33
Plane, 15, 32, 38, 48, 98, 101, 116, 162, 197, 200, 240
Planimeter, 7
Planimetric, 6, 30, 67, 101, 124, 156, 234
Plot, 148, 181
Pointers, 54, 79, 80, 129
Polar, 199, 202, 247
Polyconic, 99
Polygon, 10, 33, 49, 50, 53, 57, 80, 84, 93, 140, 147, 150, 156, 159, 164, 175, 227
Polynomial, 92, 108, 115
Postclassifier, 217
Precision, 25, 48, 58, 62, 64, 70, 86, 88, 124, 151, 164, 198, 215, 216, 222, 225,
Preprocessing, 24, 76, 118, 124, 144, 154, 157, 167, 183, 211, 223, 235, 247
Probability, 122, 169, 227
Programmers, 62, 229
Propagation, 231
Proximal, 177
Proximity, 7, 15, 57, 157, 165, 172, 231, 234
Pyramid, 42

Quadrant, 44, 205
Quadrat, 72
Quadtree, 42, 59, 242
Quality-assurance, 182
Quantization, 209
Query-by-example, 136

Radar, 123, 193, 206, 209, 231
Radiation, 195
Radiometer, 202
Raster, 33, 58, 68, 77, 88, 90, 93, 106, 113, 143, 159, 178, 184, 192, 198, 202, 214, 217, 222, 243
Rasterize, 81, 83
Reclassification, 144
Rectification, 62, 76, 93, 97, 101, 103, 154
Recursively, 38
Redundancy, 46, 53, 84, 90, 130, 140, 238
Red-green-blue, 213
Reflectance, 150, 160, 188, 193, 196, 206
Refresh, 188
Regression, 106, 168, 182, 223
Relational, 49, 53, 59, 80, 134, 147, 169
Relief, 67, 232, 251
Remote-sensing, 194
Resampling, 42, 93, 153
Resel, 12, 37
Residual, 216, 251
Rhumb, 100
Root-mean-square, 226
Rotation, 47, 88, 99, 107, 131, 199, 201
Rubber sheet, 106
Run-length, 41, 90, 91

Sample, 29, 36, 69, 72, 135
Satellite, 39, 66, 96, 124, 145, 160, 172, 183, 186, 192, 202, 211, 218, 236, 240, 242, 247, 251
Scale, 4, 6, 11, 15, 34, 44, 55, 62, 86, 89, 91, 99, 110, 119, 124, 137, 145, 153, 174, 197, 209, 212, 218, 223, 234, 237
Scan, 88, 90, 96, 151, 188, 199, 203
Scan-digitizing, 212
Security, 26, 64, 128, 137, 140, 247
Sensor, 66, 77, 123, 150, 160, 193, 199, 217
Sequential, 78, 132, 142, 147, 204
Shadow, 120, 217
Shrinkage, 86, 96
Shutter, 197
Signatures, 145, 196, 206
Significance, 122, 222, 226
Simulate, 113, 156, 209
Skeletonizing, 81, 84, 90, 215
Skewed, 199
Skewness, 167
Slivers, 94, 96
Slope, 27, 78, 114, 146, 162, 171, 217, 226, 235, 244
Small-scale, 11
Snapping, 95
Sonar, 226
Sorting, 80, 166, 177, 217, 218
Spaghetti, 58
Spatially-referenced, 3
Specifications, 63, 99, 127, 145, 201, 225
Spectrum, 80, 193, 195, 203
Standards, 49, 55, 62, 68, 133, 225, 240
Statistics, 151, 161, 167, 170, 182
Stereoplotter, 125
Stereoscopes, 124
Stratification, 75, 141, 217
Subsets, 139
Supervised, 149, 218
Survey, 2, 21, 33, 54, 65, 67, 110, 154, 222, 226, 240, 244, 247
Synchronization, 129
Synergism, 215
Syntax, 80

Tabular, 61, 123, 174, 235, 245
Tags, 127
Taxonomy, 120, 144
Television, 186, 188
Temporal, 29, 188, 195, 202, 207, 225, 236
Terrain, 4, 9, 63, 70, 114, 146, 159, 163, 177, 198, 217, 222, 231, 234, 251
Tessellation, 38
Texture, 91, 120, 161, 210, 235
Theme, 4, 32, 78, 214, 222, 224, 234
Thermal-infrared, 193
Thinning, 84, 93
Three-dimensional, 30, 32, 91, 103, 110, 117, 124, 156, 179, 232, 250
Throughput, 221
Tiles, 141, 195
Tilt, 67, 198
Timeliness, 191
Tolerance, 92, 95, 172
Tone, 119, 210
Topographic, 4, 33, 55, 67, 124, 154, 177, 200, 201, 223, 238, 247
Topology, 30, 49, 55, 80
Tract, 65, 114, 216
Trafficability, 146
Transaction, 129
Transect, 72
Transformations, 15, 106, 161, 215
Traverse, 9, 97, 137, 147, 188
Trend-surface, 115, 118
Triangular, 38, 116
Triangulate, 86
Trigonometric, 197
Tuple, 41, 134, 138, 147
Typewriter-like, 184

Ultraviolet, 193, 195
Unaligned sampling, 74
Union, 164, 246
Univariate, 170
Unsupervised, 149, 151, 213, 214
Update, 26, 57, 69, 124, 139, 215, 225, 235
Universal Transverse Mercator, 89, 99, 109

Validation, 28
Variable, 28, 79, 114, 128, 146, 160, 169, 182, 218, 235

Variable-length, 127
Variance, 150, 167, 168, 246
Vector, 48, 53, 77, 80, 88, 102, 107, 113, 129, 136, 140, 154, 163, 170, 184, 186, 215, 223, 230, 234
Verify, 70, 122, 227, 231
Vernier, 124
Video, 156, 185, 188, 190
Viewpoint, 119, 144, 153
Viewshed, 164, 167, 171
Visibility, 164, 171
Visualize, 251
Volume, 29, 32, 41, 58, 91, 133, 141, 165, 179, 183, 223, 229, 235, 247

Watershed, 156, 159, 235
Wavelength, 77, 152, 193, 196, 200
Weighting, 115, 116
Whisk-broom, 203
Window, 141, 160, 170, 244
Wireframe, 231

Yaw, 199
Yield, 181, 240, 241, 242

Zones, 99, 158, 218, 231, 233
Zoom transfer scope, 124